Radioactive Waste

Radioactive Waste
Management and Regulation

Mason Willrich

and

Richard K. Lester

with

Stephen C. Greenberg
H. Clyde Mitchell
Daniel A. Walker

THE FREE PRESS
A Division of Macmillan Publishing Co., Inc.
NEW YORK

Collier Macmillan Publishers
LONDON

The Free Press
A Division of Macmillan Publishing Co., Inc.
866 Third Avenue, New York, N.Y. 10022

Collier Macmillan Canada, Ltd.

Library of Congress Catalog Card Number: 77-80228

Printed in the United States of America

printing number
1 2 3 4 5 6 7 8 9 0

Library of Congress Cataloging in Publication Data

Willrich, Mason
Radioactive waste

Bibliography: p.
Includes index.
1. Radioactive waste disposal--United States.
I. Lester, Richard K., joint author. II. Title.
TD812.W55 363.6 77-80228
ISBN 0-02-934560-X

Contents

Preface

This book is based on a research project conducted during 1976 under the auspices of the Energy Laboratory at the Massachusetts Institute of Technology (MIT Energy Lab) for the U.S. Energy Research and Development Administration (ERDA). Mason Willrich served as principal investigator and was ably assisted by an interdisciplinary group drawn from MIT and the University of Virginia, where he is professor of law (on leave). The research group included Stephen C. Greenberg, Virginia law student; Richard K. Lester, MIT nuclear engineering graduate student (now a Visiting Research Fellow at the Rockefeller Foundation); H. Clyde Mitchell, MIT undergraduate (now a Virginia law student); and Daniel A. Walker, Virginia law student. Thereafter, Richard Lester continued to play a major role in the additional research, writing and revising which has resulted in this published product.

Our work has been reviewed by Norman C. Rasmussen, Chairman of the MIT Nuclear Engineering Department, and by David J. Rose, professor of nuclear engineering at MIT, both of whom support its conclusions and recommendations. A draft of the research report to ERDA was submitted for review and comment by the administration on September 1, 1976. The draft was also distributed to specially interested persons in other government agencies, the nuclear industry, environmental organizations, and academic institutions in order to obtain their views on its contents and the benefit of their expertise with regard to the radioactive waste problem generally. ERDA's comments and comments received from numerous individuals were taken into account in preparing this final version for publication. The work has also been revised in light of President Carter's new nuclear power policy announced on April 7, 1977. Finally, it now appears that ERDA will shortly be incorporated within a new, cabinet-level Department of Energy, as part of President Carter's energy reorganization plan. When this occurs, the references to ERDA may be assumed to signify the Department of Energy, since the substantive law pertaining to radioactive waste managment and regulation will remain unchanged.

The views expressed herein are the authors' and, other than as stated above, do not necessarily represent the views of any other person nor of any public or private institution.

May, 1977

Mason Willrich
Director for International Relations
The Rockefeller Foundation

Richard K. Lester
Visiting Research Fellow
The Rockefeller Foundation

Introduction

THE VAST BULK OF RADIOACTIVE WASTE in the United States has been and will continue to be generated by two programs: the commercial use of nuclear materials to generate electricity, and the military use of similar materials for nuclear weapons and naval propulsion. Until now, most of the radioactive waste has resulted from military activities. However, the commercial nuclear power industry has now become a major source of additional waste. Radioactive waste is thus an unavoidable consequence of our nuclear age.

Because of their intense and persistent toxicity, some types of waste must be effectively isolated from the biosphere for thousands of years. Safe management of this waste surely requires technological excellence, but the social and political issues are at least as demanding. Radioactive waste poses a major challenge to government.

The purpose of this book is to assist in developing public policy and institutions which are necessary for the safe management of radioactive waste, currently and in the long term. Indeed, an underlying hope is that our work will accelerate such development by the U.S. government.

The book focuses on the management and regulation of post-fission radioactive waste generated in the United States. This includes so-called high-level (HL) waste and low-level waste contaminated with transuranic elements, i.e., those elements—neptunium, plutonium, americium, curium, etc.—heavier than uranium. The latter waste category will be referred to as "TRU waste."

In the commercial nuclear power industry, the source of post-fission radioactive waste is the spent fuel discharged from reactors used to generate electricity. HL waste is generated during the reprocessing of spent fuel—a chemical operation in which depleted uranium and by-product plutonium may be recovered. TRU waste results primarily from reprocessing and from the fabrication of fresh fuel containing recovered plutonium.

Similarly, with regard to military uses of nuclear energy, post-fission radioactive waste originates in fuel removed from stationary reactors which produce plutonium for nuclear weapons or mobile reactors which provide propulsion for naval ships. Military HL waste emerges at a reprocessing plant; military TRU waste is generated during reprocessing and also in the manufacture of nuclear weapons.

Management of post-fission radioactive waste embraces a series of discrete but deeply interdependent operations. These operations include collection, temporary storage, treatment, packaging, transport, and permanent disposition. Regulation of waste management operations to assure safety is also composed of a number of discrete, interrelated activities. Regulatory activities include development of general criteria and specific standards, and application of established standards and criteria in licensing waste management operations. The licensing process itself includes review and approval of sites, of the design and construction of facilities, and of the conduct of operations. Regulation also encompasses the monitoring of previously licensed operations, prescription of emergency actions, and enforcement of obligations in the event of violations of legal requirements.

Post-fission HL and TRU wastes are not the only radioactive wastes which require safe management and regulation. Every step in the nuclear fuel cycle—mining and milling, uranium enrichment, fuel fabrication, reactor operation, reprocessing, and plutonium and uranium recycling—creates radioactive discharges in various forms. Depending on its toxicity, a waste stream is either actively managed or else released to the environment. Moreover, various parts of nuclear fuel cycle facilities will continue to be dangerously radioactive long after the facilities themselves are shut down, and these parts will require safe handling. We focus on HL and TRU waste streams because they raise the most pressing and difficult radioactive waste management and regulatory issues.

Although we discuss mainly the situation in the United States, important international dimensions are taken into account. Ineffective management of radioactive waste in one country may cause harmful effects in another. Moreover, low-level radioactive waste is being dumped into the ocean, and geologic disposal of HL waste beneath the ocean floor is being considered. Most of the oceans and the deep seabed are beyond the limits of national jurisdiction.

Finally, the reader should bear in mind that we consider here a narrow, though important, problem which arises in a much larger context. The specific risks posed by radioactive waste must also be compared with the risks associated with other potential environmental pollutants. Although radioactive waste constitutes a potential radiological hazard for a very long period, other possible pollutants may be even more persistent or more dangerous or both. The risks posed by radioactive waste must also be balanced against the benefits to be derived from activities which produce the waste and the consequences if those activities were stopped. The security of the United States and its allies appears to rest in part on the U.S. nuclear deterrent, and the well-being of every society depends on adequate energy. The world urgently needs practical alternatives to fossil energy, and nuclear fission has been demonstrated to be a practical way to generate electricity.

Chapter 1

Radioactive Waste

THIS CHAPTER IS INTENDED to provide readers who have not previously considered the radioactive waste problem with the information they will need to understand the analysis of waste management and regulation that follows. The discussion is written with the nontechnical reader in mind. The key terms are also explained in the glossary.

Radioactivity

Radiation is the emission and propagation of energy through matter or space. Radioactive materials emit energy in the form of electromagnetic radiation, such as gamma or X rays, or in the form of fast-moving subatomic particles such as alpha particles (the nuclei of helium atoms) and beta particles (electrons), both of which carry an electric charge, and neutrons, which carry no electric charge. As radiation penetrates matter, it interacts with its environment, and energy is transferred to the surrounding atoms, resulting in their ionization. This means that if the radiation is interacting with organic tissue, the ionized atoms may acquire new and different properties and may enter into abnormal chemical combinations, causing the decomposition or synthesis of complex molecules. The net effect is to destroy living cells and to damage the exposed tissue.

The energies of different kinds of emissions vary widely and so do their penetrating powers. For example, alpha radiation is not very penetrating compared with gamma and, to a lesser extent, beta radiation. Unlike the latter two, therefore, alpha radiation does not constitute a major hazard to man if the radiation source is external. If the source of radioactivity is inhaled or ingested, however, then all three types of radiation are important because of the close proximity of living tissue.

We are concerned about radioactive waste because it is a potential radiological hazard to man and other forms of life. A scale against which our concern can be measured is the fact that we all live in a radioactive environment.

The biosphere—that thin envelope of soil, water, and air which contains and sustains all life on earth—receives radiation from outer space and from many naturally radioactive materials found in the earth's crust. Exposure to radiation also arises from naturally occurring sources of radioactivity that are present within living tissue. In addition, since the beginning of this century, man has begun to add to this naturally occurring "background" radioactivity. Medical X rays, emissions from nuclear power plants during normal operations, and radioactive debris from nuclear weapons tests are all sources of ionizing radiation. Finally, depending on where a person lives and what he does, he is exposed to more or less radiation.

Some of the more important factors affecting the magnitude of the overall natural radiation level are altitude, geologic features, geographic location, and type of dwelling. For instance, for the first few kilometers above the earth's surface, the cosmic component of natural background radiation doubles for every 1,500-meter increase in altitude, and the natural background level increases significantly in areas of high natural uranium and thorium content.[1] The magnitude of the variations in exposure to radiation caused by local differences in the natural background often exceeds the total exposure from man-made sources. For example, a person living in Denver, Colorado, at an elevation of about one mile above sea level receives more radiation than a person who works in a normally functioning nuclear-power plant on the James River in Virginia and who lives nearby.

When a radioactive nucleus emits radiation—usually an alpha or beta particle or a gamma ray—it "decays" into another nucleus which itself may or may not be radioactive. (The nucleus, composed of protons and neutrons, is the positively charged core of an atom.) A sequence of decays through two or more radioactive nuclei is called a decay chain. Every decay chain eventually terminates with a stable, nonradioactive nucleus.

Each type of radioactive nucleus, or radioisotope, has a characteristic fractional decay rate. Of course, it does not make sense to talk about the fractional decay rate of an individual nucleus, since at any instant the nucleus in question either will or will not have decayed. But the concept becomes meaningful if, as is invariably the case, we are dealing with a large number of nuclei of the same type. In this situation we can legitimately speak of a collective fractional decay rate. For instance, the radioisotope cesium-137 decays at the rate of just over 2 per cent per year. The decay rate is independent of the absolute amount of the radioisotope present. A ton or a microgram of cesium-137 will decay at the same fractional rate.

A commonly used measure of decay rate is the radioactive half-life. This may be defined as the length of time required for half of the nuclei in a sample of a radioisotope to decay to another nuclear form. Obviously, a radioisotope with a long half-life must have a low fractional decay rate, and, conversely, one with a short half-life will have a high fractional decay rate.

TABLE 1.1. **Half-lives of Some of the Constituents of Radioactive Waste**[2]

Radionuclide		Half-life (years)
Americium	241	458
Americium	243	7,370
Cesium	135	3,000,000
Cesium	137	33
Curium	242	0.45
Curium	243	32
Curium	244	17.6
Iodine	129	17,000,000
Iodine	131	8 days
Krypton	85	10.8
Neptunium	237	2,140,000
Plutonium	239	24,400
Plutonium	241	13
Radium	226	1,600
Strontium	90	28
Technetium	99	212,000
Tritium		12.3

The half-lives of different radioisotopes span an enormous range from millionths of a second to billions of years. After a period equal to 10 half-lives, the radioactivity of a radioisotope has decreased to 0.1 percent of its original level. Table 1.1 indicates the half-lives of some of the constituents of radioactive waste.

In order to calculate the amount of radioactivity from a particular source, it is not sufficient to know only the half-life of the radioisotope involved. It is also necessary to know the quantity present. The level of radioactivity, which is the product of the quantity of a particular radioisotope and the fractional decay rate, is measured in curies. A curie is a unit which measures the absolute (as opposed to the fractional) rate of nuclear decay. One curie is defined as being equal to 3.7×10^{10} nuclear decays (or disintegrations) per second. It should be noted that this unit describes neither the energy nor the type of radiation that is emitted during a decay.

We need to know more than the number of curies in order to estimate the biological significance of a volume of material that contains one or more radioisotopes, perhaps mixed with nonradioactive materials. All radioactivity is not the same. As we have seen, different radioisotopes emit different forms of radiation at different energies as they decay.

Furthermore, different living organisms and different organs within the same organism all have varying degrees of sensitivity to radiation. Since many effects of ionizing radiation are cumulative, individual exposure histories are important. These exposure histories, together with individual radiation tolerance levels, can vary significantly. For internal sources of radiation, the physical and chemical form of the radioisotope and its route of intake are also important.

If a human being is exposed to excessive amounts of radiation, depending on the circumstances, the harmful effects may be immediate death, life shortened by radiation-induced cancer, radiation-induced genetic change which may affect subsequent generations, or temporary ill health followed by complete recovery. It is important to note that excessive exposure to harmful levels of ionizing radiation may be caused by contact with a particular radiation source or multiple contacts with many sources. The rate of absorption of radiation is also important, since damaged tissue has more chance to recover from radiation effects if the absorption takes place over a long rather than a short period.

How then do we protect ourselves from the potentially harmful effects of excessive radiation? The relation between the magnitude of radiation dose* and the resulting effect on human health is difficult to determine. The government establishes dose limits, or Maximum Permissible Doses, which are standards for radiation absorption per unit time.[3] On the basis of these limits, Recommended Concentration Guides (RCG's, formerly known as Maximum Permissible Concentrations, MPC's) have been established. RCG's provide guidance as to the concentrations of radionuclides in water suitable for unrestricted use.[4] The calculated RCG values are, therefore, a measure of the radiotoxicity of a nuclide and are frequently used in calculations of radiological hazard.

In order to calculate the radiation dose to the human population resulting from the presence of a particular radioactive source, it is necessary to predict all the routes by which radionuclides would be transported, including concentration mechanisms, from their point of release to man. In order to determine the critical exposure pathways to man, atmospheric, hydrologic, and, if the source is initially buried, geologic pathway models must be developed.

Post-fission Radioactive Waste

Radioactive waste may be defined broadly as waste material that is contaminated or possibly contaminated by radioactive isotopes. Less generally, radioactive waste is residual material that is removed from the nuclear fuel cycle and held because its release would constitute a hazard to man or to the environment. Such waste does not include radioactive emissions, within prescribed limits, which occur during the normal operations of various nuclear facilities. As we use the term, "waste" includes residual material from which certain materials might ultimately be recovered for subsequent use.

The commercial and military post-fission radioactive waste that is the subject of this work is generated by nuclear fission in power reactors

*Dose may be defined as the quantity of energy imparted to a mass of material exposed to radiation. A unit of dose is the rad, which is equal to one hundred ergs of absorbed energy per gram of absorbing material. Another dose unit is the rem (Roentgen Equivalent Man), which is the dose of any ionizing radiation that will produce the same biological effect as that produced by 1 roentgen of high voltage X-radiation (see the glossary for a definition of roentgen). It is related to the rad by a quality factor, the Relative Biological Equivalent (RBE).

which produce electricity for commercial use, and also in reactors which produce plutonium for nuclear weapons and in propulsion reactors for submarines, missile cruisers, and aircraft carriers. Post-fission radioactive waste may be differentiated into various categories such as high-level (HL), transuranic contaminated low-level (TRU), and non-TRU waste.

HL waste is composed of hundreds of radioisotopes, some in trace amounts and some in very high concentrations, some with short and some with extremely long half-lives. Various isotopes are typically gamma or beta or alpha emitters. Because of the toxicities and long half-lives of some of the radioisotopes it contains, HL waste will constitute a potential radiological hazard for hundreds, thousands, or even hundreds of thousands of years.

TRU waste contains numerous radioisotopes in much lower concentrations than HL waste. It is, however, especially troublesome because it is contaminated with transuranic elements, including plutonium, which have very long half-lives. Indeed, for a given amount of electricity generated with nuclear fuel, more than half of the plutonium ultimately discharged in all waste streams will emerge in TRU waste.[5]

Low-level radioactive waste that is not contaminated with transuranic elements is not as hazardous in the long term as TRU waste. With respect to post-fission waste, it should be recognized that dividing low-level waste into TRU and non-TRU categories, or deciding what is transuranic contaminated, is a question of judgment based on a complete interpretation of complex data. This report, as noted previously, focuses on HL and TRU wastes.

There is, unfortunately, no simple way to specify the length of time over which radioactive waste will constitute a potential hazard. Figure 1.1 shows how the potential radiological hazard of solidified HL waste changes as the waste ages. The index used in this figure to measure the radiological hazard is the number of cubic meters of water necessary to dilute one cubic meter of solidified waste to the RCG level. The RCG value in water is used (rather than that in air) because, for most disposal schemes, water-borne waste presents the most significant hazard potential. Also included in the figure, for the purposes of comparison, are the hazard indices of various uranium-bearing ore deposits.

A measure of the length of time for which the waste will remain a hazard, and for which a long-term waste management strategy must presumably be designed to be effective, might be the period for which the hazard index of the waste is greater than that of a naturally occurring uranium ore. The argument would be that, since the radioactive ore existed before primitive life began—the half-life of uranium-238, whose decay chain daughters are the major contributors to the radiotoxicity of the ore, is about 4.5 billion years—and since life has evolved in the presence of this ore, we may therefore consider a degree of radiotoxicity corresponding to this level as constituting an "acceptable" risk. Figure 1.1 shows, however, that the measure of toxicity used in this case—the hazard index—is sensitive to the type of uranium ore that is used as the reference.

Some pitchblend deposits contain uranium oxides almost completely undiluted by any other materials. Only a small fraction of the world's known uranium resources occur in such deposits, and many of them have already been mined out. On the other hand, much of the uranium that is being recovered today is extracted from deposits containing less than 0.2 percent of uranium by weight.

Diluting one cubic meter of 100 percent pitchblend mineral to the RCG level would require far more water than would the dilution of one cubic meter of sandstone containing 0.2 percent of uranium. Curve B in Figure 1.1 represents HL waste from a reprocessing plant performing with uranium and plutonium recovery efficiencies similar to present design targets. This curve shows that if pitchblend were used as the basis for measuring a hazard, the waste would reach an "acceptably" low level of radiotoxicity after about three thousand years, whereas if 0.2 percent sandstone were used it would take billions of years.

It might also be argued that the use of such a hazard index to establish the toxic lifetime of the waste fails to take adequate account of the fact that in all schemes for disposal in geologic formations the rock formation itself would act as a dilutant for the waste. For example, one estimate for disposal of waste in salt beds is that the salt and associated shale would provide a dilution factor of 37,000.[7] On the other hand, the waste would not be distributed homogeneously throughout the salt, but rather would be placed in canisters dispersed at intervals throughout the formation.

An alternative way to measure the period of radiological hazard would be to determine the age of the waste when the total volume of water required to dilute one kilogram of waste to the RCG level would be equal to the volume of water required to dilute to the RCG level the mass of ore that had had to be mined to produce the quantity of nuclear

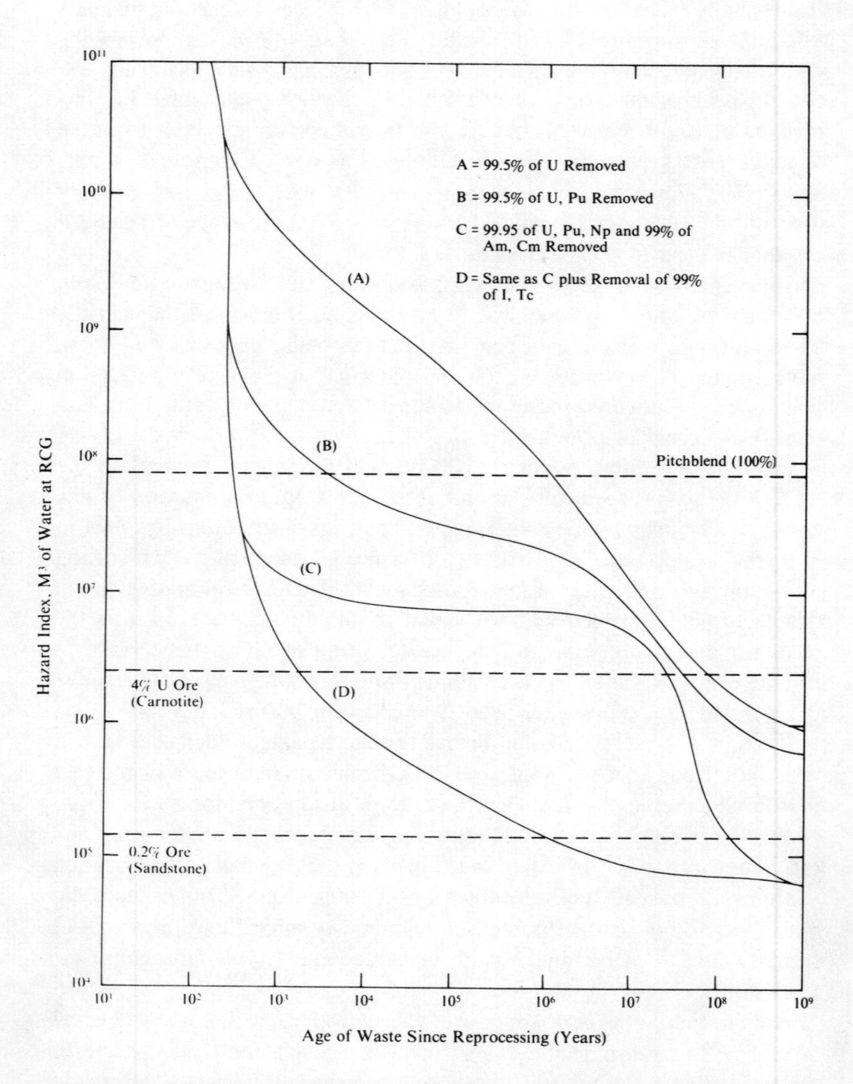

FIGURE 1.1: **Solidified high level waste hazard for different isotope removal schemes.**

fuel which, when irradiated and reprocessed, generated the kilogram of HL waste. (For light water reactors (LWRs)—the type predominantly in use in the U.S.—the amount of ore in question, assuming it contained 0.17 percent by weight of uranium, would be nearly thirty tons.) The concept is that use of nuclear fuel would ultimately result in the transfer of radioactivity from one part of the earth's crust (where it is in the form of uranium ore) to another—a waste repository. When the toxicity of the waste falls to the level of toxicity of the naturally occurring material that had been consumed to produce it, then the waste is no longer considered a hazard. Again, there are problems with this approach. The risk presented by radionuclides distributed and diluted throughout a large body of ore is perhaps qualitatively different from the risk presented by radionuclides concentrated in a canister of waste, even if the same volume of water is required to dilute the ore body and the waste canister to the RCG level. Furthermore, much of the initial radioactivity in the ore is not, in fact, consumed, but instead remains in the ore "tailings"—the waste stream generated during uranium extraction.

The previous discussion illustrates, at least in part, some of the difficulties involved in quantifying the length of time that must elapse before the waste can be considered no longer hazardous. The numerous variations on the theme of waste hazard "potential" give answers ranging upwards from hundreds of years.

Radioactive Waste Generation

Post-fission waste originates from spent fuel assemblies that are routinely discharged from nuclear power, production or propulsion reactors. The irradiated fuel rods in these assemblies contain fission products; uranium, whose original uranium-235 content has been reduced by fission; and plutonium, which is produced by neutron capture in uranium-238.

After discharge from the reactor, the materials constituting spent fuel assemblies move through a sequence of operations. Various categories of radioactive waste emerge at different points in the sequence. The operations are outlined briefly below. The radioactive waste management implications are discussed in more detail in Chapter 2.

Temporary Spent Fuel Storage

Spent fuel assemblies are temporarily stored in water-filled storage basins located near the reactor where irradiation has occurred. Radioactive decay proceeds within the stored spent fuel sufficiently to permit safe handling and safe transport. No plant for reprocessing spent fuel from commercial power reactors is presently operating in the United States. Consequently, the available storage capacity of existing basins at operating power reactors is rapidly being filled. As a result, additional temporary storage capacity, either at various reactor sites or at a central location, will be required soon.

Spent Fuel Transport

After the cooling period, commercial spent fuel rods, which are still highly radioactive, are placed in massive shielded containers for shipping from the storage basin at the reactor site to a reprocessing plant. In the case of military plutonium production, the production reactor and reprocessing plant are located on the same site so that there is no need for off-site transportation for reprocessing.

Reprocessing

Following arrival at a reprocessing plant, the assemblies are disassembled and the spent fuel rods are chopped up. The residual mixture of fuel materials and fission products is dissolved and sent through the plant, where a large number of chemical separations are performed. Plutonium and uranium are recovered.

In the commercial nuclear power industry, the uranium can be reenriched in the uranium-235 isotope, or it can be mixed with plutonium. In either case, uranium would be recycled in power reactor fuel. The recovered plutonium can be blended with uranium in mixed oxide fuel for recycling in existing light water reactors, or it can be stored for later use as fuel in breeder reactors. It is also possible to leave the plutonium in the radioactive waste, rather than extract it.

In the weapons program, the recovered plutonium is used in the manufacture of nuclear warheads. In the naval propulsion program, the uranium is recovered and reenriched, but the relatively small amounts of plutonium created during fuel irradiation are disposed of with the waste products.

Most of the radioactivity that was initially contained in the spent fuel is discharged from the reprocessing plant in the liquid HL waste stream. This waste contains almost all of the nonvolatile fission products, a small residual percentage of uranium and plutonium, and practically all of the other transuranic elements. Other reprocessing plant wastes include: (1) low-level waste—liquid and solid—containing low concentrations of short-lived fission products and no more than trace amounts of long-lived transuranic elements; and (2) waste—cladding hulls, failed equipment, spent process materials, trash, etc.—with lower levels of radioactivity than HL waste, but with appreciable amounts of the transuranic elements. Waste volumes in these two categories are much larger than in the case of HL waste. Waste in the second category is what we have previously called TRU waste.

One plant for reprocessing commercial power reactor fuels was previously in operation at West Valley, New York. The plant was shut down in 1972 for modifications and expansion. The owner, Nuclear Fuel Services, Inc., recently announced that it would not be reopened owing to the high cost of the alterations.[8] Another larger commercial reprocessing plant is currently under construction by Allied General Nuclear Services at Barnwell, South Carolina. The completion of this plant is, however, dependent on obtaining financial assistance from the federal government and a favorable resolution of pending regulatory decisions.

Plants for reprocessing production reactor fuel for weapons are located at Savannah River, South Carolina, and Richland, Washington. The reprocessing of naval reactor fuel and fuel from ERDA research and test reactors takes place at Idaho Falls, Idaho.

Temporary Waste Storage

HL waste emerges in liquid form in the course of reprocessing, primarily at the first stage of plutonium and uranium extraction. The liquid is stored temporarily in tanks at the reprocessing plant.

TRU waste emerges in various forms—liquid, and combustible and noncombustible solids—at the reprocessing plant and also at other points in the nuclear fuel cycle. The main sources of TRU waste are: the fuel cladding hulls and assembly structures from the head end of the reprocessing plant; the residuals left from the conversion of liquid plutonium nitrate to plutonium oxide powder (commercial use) or to plutonium metal (weapons use); and spent process material, general trash, and failed equipment left from reprocessing and also from commercial mixed oxide fuel fabrication and military weapon component fabrication. TRU waste is collected, temporarily stored, and sometimes mixed with other less toxic wastes at these various fuel cycle stages.

Waste Treatment

Depending on its composition, radioactive waste is treated in various ways: gases may be dissolved in liquids or adsorbed on solids, liquids may be solidified, and solids may be incinerated (resulting in gas and less solid). In general, treatment is designed to make the waste easier or safer to handle subsequently. However, every treatment method has its own set of costs and risks and almost always results in the generation of secondary waste streams.

Waste Transport

Following treatment, radioactive waste is transported to a location where it is more or less permanently disposed of. The importance of transport depends on the site of permanent disposition. No transportation is involved if HL waste is solidified in the bottom of a temporary storage tank and left there indefinitely. Long-distance transport involving railroad or truck and ship would be necessary for either seabed or ice-sheet disposal.

Permanent Disposition

A wide variety of methods for permanent disposition of HL waste is being considered. The most practical appear to be those involving emplacement in suitable geologic formations, either deep underground or under the ocean floor. Other methods under study are ice-sheet disposal and waste partitioning combined with nuclear transmutation or extraterrestrial elimination of the partitioned long-lived actinides.

No method for permanent HL waste disposition has been adopted yet. However, much military HL waste which was originally stored temporarily as a liquid in tanks has been subsequently solidified *in situ*. It now appears that the task of exhuming these solids will be very difficult and costly, and for this reason the waste may be left there for a substantial period—perhaps indefinitely.

The technology has not yet been developed for the treatment and packaging of military HL waste that has been already solidified in the storage tanks. Estimates of the cost of preparing this waste for permanent disposition range up to $20 billion. If, as some critics charge, the present in-tank storage method for solidified salt cake is continued indefinitely, then such large expenditures would be avoided. In this situation, however, the waste confinement system would be less rigorous than the methods currently under consideration for isolation of commercial high-level waste.

Some TRU waste has been disposed of, essentially irretrievably, in ocean dumping grounds. Other waste in this category has been deposited in shallow burial grounds. Not all of these land burials will permit retrieval at reasonable cost. The question of the permanent disposition of TRU waste to be generated in the future is unresolved.

Radioactive Waste Quantities

The following is a brief review of the amounts of radioactive wastes in various categories that have already been generated in the United States, together with some predictions of how these amounts may increase over the next twenty to thirty years. The discussion compares military and commercial waste quantities.

High-level Waste

As a result of military activities from the middle 1940s to the present, the former U.S. Atomic Energy Commission (AEC) and its successor, the Energy Research and Development Administration (ERDA),* have generated about 215 million gallons of liquid HL wastes.† Solidification programs have been in operation for some years at the three sites at which this waste is stored: the Hanford reservation near Richland, Washington, the Savannah River Plant in South Carolina, and the Idaho National Engineering Laboratory, near Idaho Falls in Idaho. As a result, over 80 percent of the original liquid has been solidified. The volume reduction associated with solidification has meant that, as of January 1, 1976, there was an HL waste inventory of about 75 million gallons, half of which was in solid form.[12]

More than 70 percent of this waste is currently stored at Richland, about 25 percent at Savannah River, and 3 percent at Idaho Falls.[13] It is estimated that by the early 1980s, when most of the waste will have been solidified, there will be a total of nearly 500,000 tons of residual HL solids at these three sites.[14]

In contrast, the commercial nuclear power industry has until now produced less than 600,000 gallons of HL waste. All of this waste is still in liquid form and is stored at the site of the Nuclear Fuel Services reprocessing plant at West Valley, New York. (Although 600,000 gallons of liquid HL waste are stored at the Nuclear Fuel Services site, much of this was generated during the reprocessing at West Valley of spent fuel from an AEC production reactor at Hanford. The amount attributable to the AEC cannot strictly be categorized as ''commercial'' waste.[15])

The future production rate of military HL waste is uncertain, since predictions must be based partly upon an assumed plutonium demand for nuclear weapons. Predictions of future generation rates of commercial HL waste must be based on assumed nuclear power growth rates and are therefore also subject to considerable uncertainty. It has been estimated, however, that the commercial power industry in the United States will have generated 60 million gallons of liquid HL waste by the year 2000,[16] and that not until about 2020 will the commercial power industry have produced the volume of liquid HL waste that has already

*With the passage of the Energy Reorganization Act of 1974 the former AEC was fissioned into two parts—the Nuclear Regulatory Commission (NRC) and the ERDA (nuclear and nonnuclear energy research, development and demonstration, enrichment services, and nuclear weapon materials production).

†Up to 1974, some 205 million gallons had been generated.[9] During that year, it is estimated that 7.5 million gallons were added,[10] and in 1975 a further 4 million gallons of HL waste was generated.[11]

been produced by U.S. military programs.[17] More recent estimates, perhaps based on a more conservative industry growth rate, indicate that current volumes of military wastes are more than ten times the amount that will be accumulated by the nuclear power industry by the year 2000.[18]

According to recent calculations, the volume of all the solidified HL waste produced by the commercial nuclear power industry through the year 2000 will be equivalent to a cube 66 feet on each side,[19] whereas the equivalent cube for the existing military HL waste would have a side measuring approximately 200 feet.[20] Of course, this should not be interpreted to mean that these small volumes are all that will be required to store the waste. Much larger storage volumes will be necessary. But the calculation does indicate that the main problem in safely managing HL waste will be one of confining the radioactivity, rather than of finding enough storage space.

A volumetric comparison alone is inadequate when comparing military and commercial wastes. Owing to differences in treatment processes, one ton of spent fuel from a plutonium production reactor produces a larger volume of liquid HL waste than would be produced by one ton of spent commercial power reactor fuel. Furthermore, fuel in plutonium production reactors only receives about one-tenth as much irradiation exposure before discharge as commercial power reactor fuel,[21] and the amount of radioactivity in spent production fuel is therefore correspondingly lower.

Both of these factors mean that the radioactive inventory of a unit volume of military liquid HL waste is very much less than that of the same volume of commercial liquid HL waste. This difference is maintained in the final, solid volumes. It has been estimated that the long-lived fission product concentration in the solidified military waste is on the order of a thousand times less than in commercial waste.[22]

About 600 million curies of cesium-137 and strontium-90 are stored at Savannah River and Richland.[23] (About 150 million curies of this is stored separately as encapsulated solids at Richland, having been partitioned from the liquid HL waste.[24]) Projections of waste from assumed commercial processing of power reactor fuel indicate that about 700 million curies of cesium-137 and strontium-90 would be in the waste delivered to a federal repository by 1990.[25]

The conclusion that clearly emerges is that the quantity of military HL waste exceeds the amount generated by the commercial nuclear power industry, and, at least on a volumetric basis, the difference is very large. Furthermore, this difference is likely to continue at least until the end of the century.

TRU Waste

Through June, 1974, 42 million cubic feet of military low-level waste had been buried at ERDA's land burial sites.[26] Through 1973, over 9 million cubic feet of commercial low-level waste had been buried at six commercially licensed sites.[27]

Until April, 1970, no distinction was made between TRU and non-TRU low-level waste for burial purposes. At that time, the AEC decided that radioactive waste with known or detectable contamination of transuranic nuclides which would be subsequently delivered to federal burial grounds should be segregated from other types of waste and stored in retrievable form. This requires packaging and storage so that it remains readily retrievable in contamination-free containers for an interim period of twenty years, and retrievability should continue to be possible beyond that period.[28]

About 952 kilograms of plutonium are contained in the TRU waste buried at the five major ERDA burial sites.[29] Of this amount, 740 kilograms were buried before 1970 with no provision for retrieval, and the remaining 212 kilograms have been stored retrievably.[30] Through 1973, TRU waste containing some 80 kilograms of plutonium had been buried at the six commercial sites.[31] There is likely to be no further shallow land burial of commercial TRU waste if proposed regulatory changes are adopted. All future military and commercial TRU waste (and that which is stored presently) will probably be disposed of more carefully.

Military activities are currently generating low-level solid waste at the rate of 1.3 million cubic feet per year.[32] It is expected that this rate will gradually decrease in the future. Recent projections of commercial waste generation indicate that by the year 2000 there will be over 50 million cubic feet of commercial TRU waste, including about 1 million cubic feet of spent fuel cladding waste.[33] (In these estimates, no allowance has been made for volume reduction by compaction or other methods.)

In addition to the above quantities, low-level waste, military and commercial, has been placed in canisters and discharged into the sea at U.S. dumping sites off the Atlantic and Pacific coasts. Part of this waste is contaminated with transuranics. Ocean dumping of U.S. low-level waste was suspended in 1970. However, other countries are continuing this practice.

The existing amount of military TRU waste thus exceeds the current amount of commercial waste, but the dominance is not as great as in the case of HL waste.

Radioactive Waste Release

AEC, ERDA's predecessor as manager of military waste, intentionally discharged or accidentally released large quantities of HL and TRU waste. Between 1956 and 1958, about 31 million gallons of radioactive waste containing 1.3 million curies (excluding strontium and cesium, but including plutonium) were poured into soil on the Hanford reservation.[34] During the past two decades, the radioactivity has decayed to a level of about 20,000 curies.[35] In addition, the Hanford Plutonium Finishing Plant has discharged liquid effluent containing large quantities of plutonium to subsurface trenches that are not isolated from the soil. Indeed, in the case of one particular trench, it is estimated that about 40 kilograms of plutonium were released in this way,[36] creating a potentially severe radiological hazard and leading to the addition of a neutron poison to the trench in order to eliminate the possibility of a spontaneous fission reaction.* Since the beginning of operations at the Hanford site, up to 43 million curies of beta emitters and 192 kilograms of plutonium have been carried into the soil in 140 billion gallons of liquid effluents.[38] The current beta inventory in the soil is approximately 10,000 curies, about one-half of which is cesium-137 and strontium-

*Initially it was believed that this possibility was quite high, but later calculations showed that the chance of a chain reaction was in fact extremely small, even without the addition of the poison. Nevertheless, for purposes of comparison, it is noteworthy that the HL waste generated by a typical U.S. power reactor (i.e., a 1000 Megawatt electric (MWe), light water reactor) during its entire operating lifetime (about thirty years) would contain about 30 kilograms of plutonium. In this context, it should be remembered that shallow land burial of roughly 740 kilograms of plutonium has taken place at ERDA's five principal burial sites without provision for retrievability.[37]

90.[39] The transport of most of the radionuclides, including cesium-137, strontium-90, and plutonium, has been significantly retarded by the soil.[40]

At Savannah River and the Hanford reservation, liquid HL waste is neutralized and stored in carbon steel tanks. The liquid HL waste tank farms at the Hanford reservation have developed twenty leaks, causing losses amounting to about 450,000 gallons of neutralized waste.[41] All the waste tanks which have leaked were constructed prior to 1956. The largest leak so far occurred in 1973 and caused the escape of 115,000 gallons of HL waste containing 50,000 curies[42] into the surrounding soil. Eight leaks have occurred at the Savannah River Plant, but only one, of about 100 gallons, is known to have resulted in contamination of the surrounding soil.[43] These leaks have neither killed nor injured anyone to date. Nonetheless their hazard will remain for hundreds or thousands of years. At Idaho Falls, where the liquid HL waste is unneutralized and stored as acid in stainless steel tanks before solidification, there have been no leaks.

The HL waste management record has improved over the years. For example, at the Hanford reservation the controlled releases of radioactive liquids to the environment have been significantly reduced: in 1965 about 140,000 curies of beta activity were intentionally discharged, but in 1970, as a result of operational and design changes, the amount had dropped to 13,000 curies.[44] Efforts to upgrade performance continue.

Potential Causes of Harmful Effects

Many events may cause the release of radioactivity from waste management activities. They can be classified as accidents due to human failure, intentional acts, or natural phenomena.

Accidents may be due to either inadequate construction or improper operation. An example of the first would be the leaks which occurred in carbon steel, military HL waste tanks.

Intentional acts against waste management facilities may be characterized as acts of either warfare or sabotage. For instance, the temporary HL waste tanks adjacent to a reprocessing plant might be sabotaged for the purpose of achieving long-term local contamination. Surface waste management activities might be targeted with nuclear weapons to increase the degree, duration, and coverage of contamination after a nuclear war. However, the same weapons would be much more effective if they were used initially in the destruction of cities or military targets.

Finally, various natural events can abruptly or chronically initiate releases of radioactive materials to the environment. Abrupt events include floods, earthquakes, volcanoes, and, conceivably, meteors. They may result in immediate environmental contamination, or they may expose the final waste form to slow but continual degradation and subsequent seepage into the biologically active media of soil and water. In chronic release scenarios, the essential question is how much waste will reach the soil or watershed during its period of toxicity. Chronic events include the possible shifts of water tables and river courses, the glacial erosion caused by ice ages, slow but continual geologic faulting, leaching processes, and surface ice flows and underground lake migration in Antarctica.

Conclusions

From the preceding outline of the basic facts of the post-fission radioactive waste problem in the United States, several conclusions may be drawn:

1. Radioactive wastes must be safely managed to prevent possible radiological hazards to life.
2. As a consequence largely of nuclear weapons programs, the U.S. government has already committed itself to a radioactive waste management responsibility of major proportions which is still growing.
3. Because of the maturing nuclear power industry, the United States is generating a rapidly increasing volume of radioactive waste. The vast bulk of this waste remains in the form of spent fuel elements in temporary storage basins at commercial power reactor sites. Whether or not these elements are reprocessed, they pose a growing waste management problem.
4. In the past, the U.S. government's record of management has been marred in a sufficient number of instances to be a cause of concern.

An inevitable legacy of our nuclear age, radioactive waste constitutes a potential Nth century hazard.

Notes to Chapter 1.

1. Unruh, C. M., "Nuclear Radiation," *Nuclear Technology,* 24 (December 1974), 314–322.

2. Table data extracted from: El-Wakil, M. M., *Nuclear Heat Transport,* International Textbook Company, Scranton, 1971, 430.

3. See *Standards for Protection Against Radiation,* 10 C.F.R.20, Superintendent of Documents, U.S. Government Printing Office, Washington, D.C., 1973.

4. *Ibid.*

5. Bray, G. R., "Other-Than-High-Level Waste," in *Proceedings of Nuclear Regulatory Commission Workshop on the Management of Radioactive Waste: Waste Partitioning as an Alternative,* held at the Battelle-Seattle Research Center, Seattle, Wash., June 8–10, 1976, NR-CONF-001, U.S. Nuclear Regulatory Commission, Washington, D.C., 1976, 102.

6. With the exception of the uranium ore reference lines, all the information in this figure was taken from: Dance, K. D., *High-Level Radioactive Waste Management: Past Experience, Future Risks, and Present Decisions,* Report to the Resources and Environment Division of the Ford Foundation, April 1, 1975, 21.

7. Bell, M. J., *Heavy Element Composition of Spent Power Reactor Fuels,* ORNL-TM-2897, Oak Ridge National Laboratory, Oak Ridge, Tenn., May 1970.

8. *Nucleonics Week,* Vol. 17, No. 39 (September 23, 1976), 1.

9. U.S. General Accounting Office, *Isolating High-Level Radioactive Waste from the Environment: Achievements, Problems, and Uncertainties,* Report to the Congress, B-164052, Washington, D.C., December 18, 1974, 3.

10. *Ibid.*

11. *Hearings Before the Subcommittee on Legislation of the Joint Committee on Atomic Energy,* 94th Cong., 2d Sess., February 4, 1976, at p. 1460.

12. *Ibid.*

13. Dance, K. D., *High-Level Radioactive Waste Management,* 27.

14. The volume of the current inventory of military wastes, if solidified as salt cake or calcine, would be about 220,000 m^3 (see footnote 22). The density of salt cake is roughly 1,800 kg/m^3 (WASH-1528, *Environmental State-*

ment: Future High-Level Waste Facilities, Savannah River Plant, Aiken, S.C., April 1973, 16.) The mass of the current inventory of military waste, in the form of salt cake, would be 435,000 short tons.

15. See note 8.

16. GAO (*Supra* note 9), 3.

17. Dance, K. D., (*Supra* note 6), 15.

18. February hearings (*Supra* note 11), p. 1464.

19. U.S. Energy Research and Development Administration, *Alternatives for Managing Wastes from Reactors and Post-Fission Operations in the LWR Fuel Cycle,* ERDA-76-43, Washington, D.C., May 1976, Vol. 1, 3.12.

20. *Nucleonics Week,* Vol. 17, No. 30 (July 27, 1976), 13.

21. U.S. Energy Research and Development Administration, *Final Environmental Statement, Waste Management Operations, Hanford Reservation, Richland, Washington,* ERDA-1538, Washington, D.C., December 1975, Vol. 1, X-33.

22. Letter of October 27, 1976, from Frank Baranowski, Division of Nuclear Fuel Cycle and Production, ERDA, to Mason Willrich.

23. *Ibid.*

24. *Ibid.*

25. *Ibid.*

26. U.S. General Accounting Office, *Improvements Needed in the Land Disposal of Radioactive Wastes—A Problem of Centuries,* Report to the Congress, RED-76-54, Washington, D. C., January 1976, 4–5.

27. *Ibid.*

28. *Ibid.*

29. *Ibid.*

30. *Ibid.*

31. *Ibid.*

32. *Ibid.*

33. ERDA-76-43 (*Supra* note 19), 3.15–3.22.

34. GAO (*Supra* note 9), 14.

35. *Ibid.*

36. Smith, A. E. (comp), *Nuclear Reactivity Evaluations of 216-Z-9 Enclosed Trench,* ARH-2915, Richland, Wash., December 1973.

37. GAO (*Supra* note 26), 5.

38. Baranowski, F. (*Supra* note 22).

39. *Ibid.*

40. *Ibid.*

41. U.S. Nuclear Regulatory Commission, *Alternative Processes for Managing Existing Commercial High-Level Waste,* NUREG-0043, Battelle-Pacific Northwest Labs, Richland, Wash., April 1976, 38.

42. *241-7-106 Tank Leak Investigation,* ARH-2874, Atlantic Richfield Hanford Co. (ARHCO), Richland, Wash., November 1973.

43. Dance, K. D., (*Supra* note 6), 28.

44. Baranowski, F. (*Supra* note 22).

Chapter 2

Radioactive Waste Management

IN CHAPTER 1 WE CONCLUDED that post-fission radioactive waste is a growing problem with both immediate and long-term ramifications. The problem requires enlightened, farsighted management.

This chapter, therefore, focuses on radioactive waste management. It does so in a strategic, not a tactical sense. We consider key decisions, rather than day-to-day operations. Of course, management involves a decisional process that carries the radioactive waste problem forward from the present into the future. As we shall see, radioactive waste management is a paradigm of decision making under uncertainty. It is a task, moreover, that is strongly characterized by multiple conflicting objectives and the distribution of risks and costs through time.

The purpose of this chapter is to present in a coherent manner the strategies that are under consideration for the management of post-fission radioactive waste. The discussion is structured to provide readers with an understanding of how the sequences of key decisions that constitute waste management strategies are interrelated. It is important to determine the constraints each decision imposes on the available technological options. It is also important to consider the risks, costs, and benefits associated with the various strategies. Our task is to identify the strategies and the major issues which they raise. We do not attempt quantification. Therefore, the analysis which follows is intended to provide a framework for the problem that is sufficiently flexible to cope with the uncertainties involved, yet rigid enough to provide guidance for waste management decision making.

Overview

The main waste management strategies currently available for commercial and military HL and TRU wastes are shown in Figures 2.1 and 2.2. The strategic decisions have been highlighted. The sequences of decisions correspond chronologically to the actual material flows (for example, from reactor spent fuel discharge to permanent disposal of HL waste). This is not necessarily the order in which the strategic decisions will be or should be made. The decisional interdependence of waste management planning, unlike the flow of the waste itself, does not move in one direction through time. For example, the selection of a method of permanent disposition may influence the selection of the form of solidified waste and possibly the requirements for waste composition. Such decisional interdependence is a recurrent theme.

Military and commercial management are separated. In the short term, the technologies required to deal with these wastes of different origin are significantly different, particularly in the high-level category. Furthermore, important early decisions in the case of military waste, starting with the initial decision to reprocess, have already been made, whereas almost all of the major issues concerning commercial waste generation and management (with the notable exception of the decision to develop a substantial nuclear electric generating capacity) have yet to be settled.

Because of the future importance of nuclear power and the unresolved nature of the accompanying post-fission waste management decisions, it will be illuminating to consider in the following order the general implications for nuclear power development of reprocessing, the waste management implications of a decision not to reprocess, and the radioactive waste implications of reprocessing. The stage will then have been set for a discussion of the management of commercial HL, military HL, commercial TRU, and military TRU wastes.

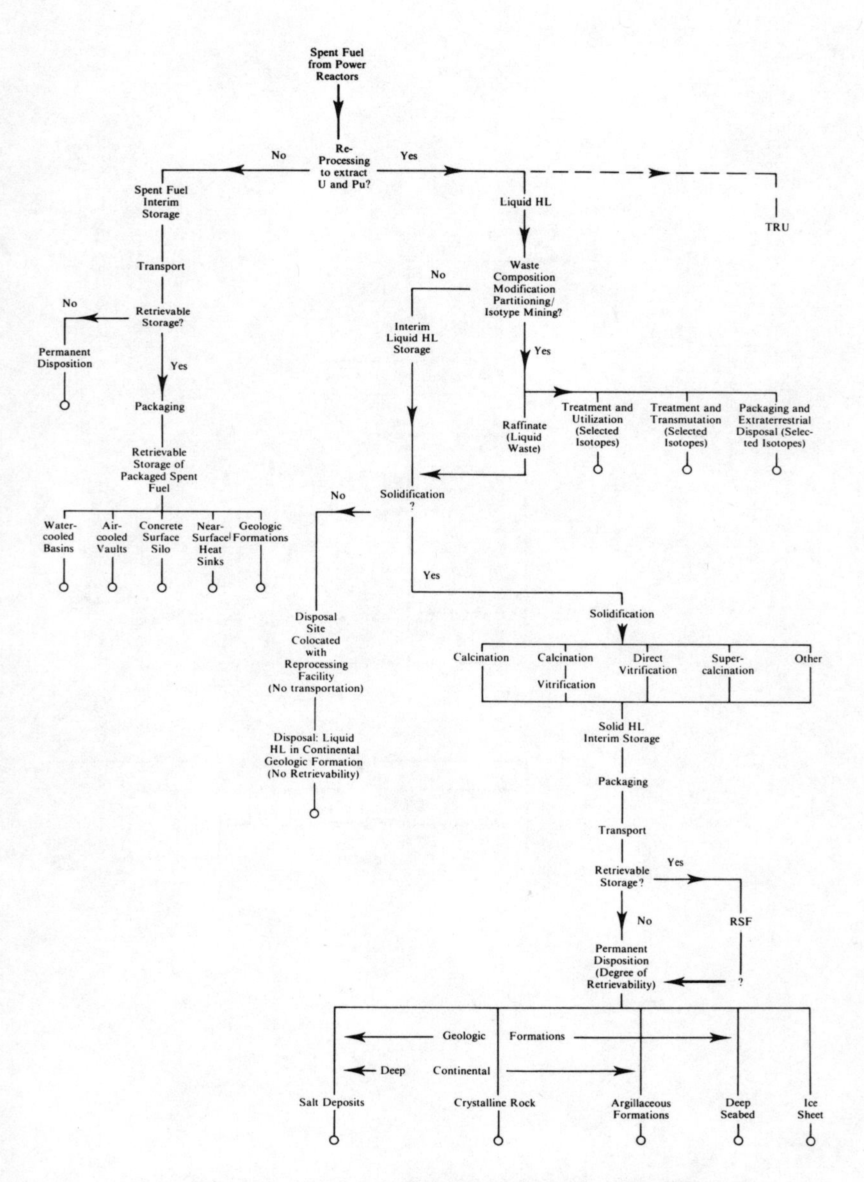

FIGURE 2.1: **Commercial HL waste management strategies.**

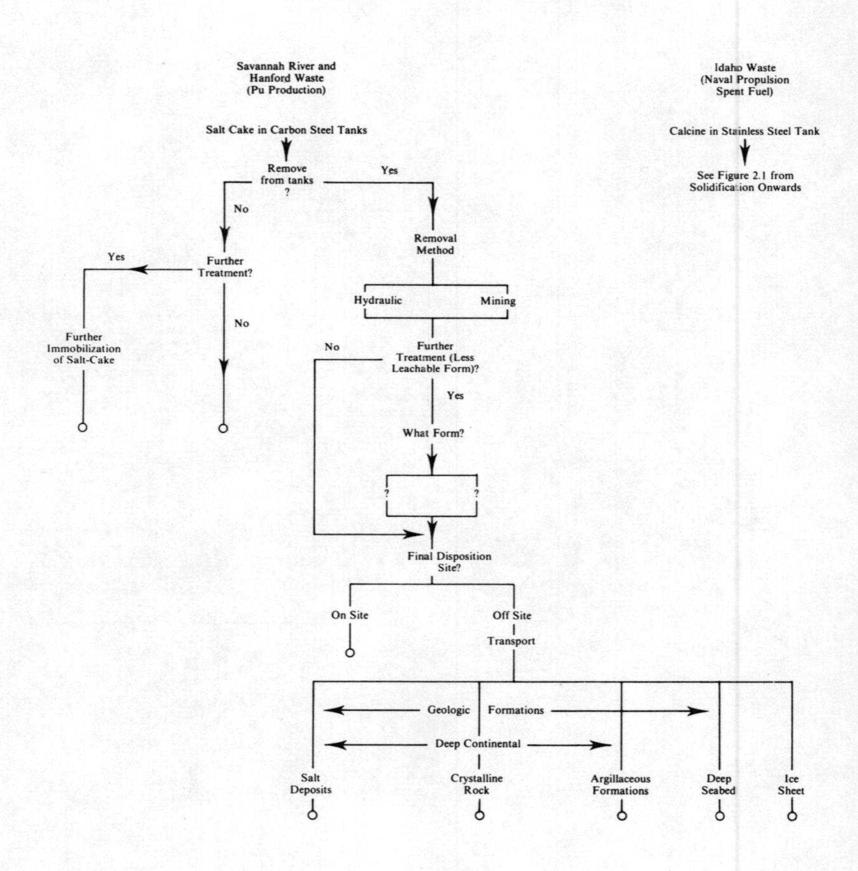

FIGURE 2.2: **Accumulated military wastes management strategies.**

Reprocessing: Implications for Nuclear Power Development

The first decision in Figure 2.1 is whether to reprocess spent fuel from commercial power reactors. The reprocessing decision profoundly affects the whole future development of nuclear power. Accordingly, waste management issues must be kept in perspective, and neither exaggerated nor lost from view. President Carter announced on April 7, 1977 a new policy to "defer indefinitely the commercial reprocessing and recycling of the plutonium produced in the U.S. nuclear power programs." This decision was premised on the increased risks of nuclear weapon proliferation that would result from the development of nuclear power in a way which would involve widespread direct access to plutonium or high-enriched uranium—materials usable in nuclear explosives. At this writing it remains to be seen what effect the Carter decision regarding commercial reprocessing will have on the development of nuclear power in other countries.

The short-run incentive for reprocessing spent power reactor fuel is to extract uranium and plutonium so that these fissionable materials can be recycled as fuel in commercial power reactors of the type now operating. It is not clear that the value of the fissionable material extracted from spent fuel will outweigh the cost of reprocessing. Uncertainty arises from a number of sources.

First, the value of plutonium and uranium recovered through reprocessing depends on the price of natural uranium. At the present time, the extent of uranium ore resources is not well known. Although the selling price of uranium ore has been increasing rapidly, future price trends are highly uncertain and are likely to be affected by other factors in addition to the availability of uranium ore.

The second major source of uncertainty is the cost of spent fuel reprocessing and the cost of fabricating fuel containing mixed plutonium and uranium oxides. Mixed oxide fuel fabrication is much more expensive than the fabrication of the low-enriched uranium fuel used currently in light water reactors. Two types of uncertainty are associated with these costs: (1) Neither reprocessing nor mixed-oxide fuel fabrication has reached the stage at which it can be called a "mature" industry. Cost estimates for commercial plants have to be made by extrapolation from smaller scale demonstration-type plants; (2) The regulatory requirements for these plants are in a process of substantial change. In particular, the safeguards against theft or diversion that must accompany plutonium through all stages of recycle are undetermined.

Therefore, the calculations often used to estimate the economic feasibility of spent fuel reprocessing and plutonium and uranium recycle cannot provide a conclusive answer to the question of whether or not to reprocess.

To put the decision in perspective, it is necessary to include its long-term implications. Many of the important implications for waste management will be dealt with separately, however, and the implications for the future of nuclear power are merely outlined here.

A decision not to reprocess spent fuel, if adhered to, is likely to mean that nuclear power will only play a short-term, interim role in meeting U.S. energy demands. The duration of that role would be limited by the extent and quality of uranium ore deposits known and yet to be discovered.*

The present generation of commercial power reactors, the light water reactors (LWR's), use slightly enriched uranium fuel that contains 2 to 4 percent uranium-235 and the remainder uranium-238. Fission of uranium-235 in present LWR's is accompanied by the conversion of some of the uranium-238 isotope into plutonium, which is easily fissioned, but LWR's are relatively inefficient converters. Without reprocessing, therefore, fission is limited to roughly 70 percent of the uranium-235 contained in fresh fuel—equivalent to about 2 percent of all the uranium loaded into the reactor—and that fraction of the plutonium produced from the initial inventory of uranium-238 which fissions before the depletion of the fissile inventory and the build-up of neutron-absorbing "poisons" render the core incapable of sustaining a chain reaction and fuel must be discharged. The exact percentages vary depending on which particular fuel management scheme is chosen.

If the decision is made to reprocess spent fuel and to allow the recycle of uranium, but not plutonium, then the uranium resource constraint

*However, in the unlikely event that it becomes technically feasible, economically attractive, and environmentally acceptable to extract the uranium that is known to exist in very low concentrations in enormous quantities in certain shale and granite formations, and in seawater in much lower concentrations still, the duration of a nuclear power industry without spent fuel reprocessing would no longer be limited to the next few decades.

will still basically apply. The recycled uranium would add at most about 20 percent to the energy that could ultimately be produced. If, however, the decision is to permit spent fuel reprocessing and to allow recycling of both uranium and plutonium in existing LWR's, the recycled plutonium would add perhaps an additional 25 percent to the energy output from nuclear power.

The extension to the lifetime of an LWR economy gained by a decision to recycle plutonium would depend on the scale and growth rate of the industry. For instance, if there were enough uranium to support an LWR population of constant size for forty years without recycling plutonium, then a decision to recycle would add about a decade to the life of the industry. In a growing LWR economy, however, a 25 percent increase in the available energy resource would produce less than a 25 percent increase in the lifetime of the industry. Thus, even with large-scale reprocessing and recycling of uranium and plutonium, a nuclear power industry relying solely on the existing type of commercial nuclear reactors would remain an interim energy option, given the current resource constraints.

The future of nuclear power as a long-term energy option hinges primarily on the successful development and commercialization of a plutonium breeder reactor, and such a breeder necessarily requires large-scale reprocessing of spent fuel. Plutonium breeder reactors use mixed plutonium–uranium fuel surrounded by a uranium blanket. Through uranium-238 neutron capture and subsequent conversion, fission of plutonium in a breeder reactor results in the production of more plutonium than is consumed. In short, a breeder reactor is a much more efficient converter of plentiful uranium-238 into plutonium than is a light water reactor.

A plutonium breeder of the liquid metal cooled, fast neutron reactor type (LMFBR) is the long-term technological option currently being pursued as the highest priority by every country with a major nuclear power program, except possibly for the U.S. The breeder reactor development programs in a number of countries including France, Great Britain, the Federal Republic of Germany, and the Soviet Union are currently on more advanced schedules than is the U.S. program, and the most likely impact of President Carter's nuclear power policy will be to widen this apparent gap in development schedules.

It is possible to construct a strategy in which spent fuel is reprocessed and the plutonium is not recycled to existing LWR's, but rather is stored for an interim period and used subsequently in plutonium breeder reactors. This may well occur in Europe and the Soviet Union because of the earlier introduction of breeders into commercial use.

On April 7, 1977, President Carter announced his decision to "restructure the U.S. breeder reactor program" and to defer the date when breeder reactors would come into commercial use in the U.S. It was further decided to defer indefinitely the Clinch River LMFBR demonstration project and to cancel all commercialization, component procurement, and licensing efforts for the project. The outcome of this decision depends ultimately on Congressional authorization, and at this writing there are substantial efforts underway in the Congress to keep the Clinch River project alive.

Like the decision to defer commercial reprocessing, the Carter administration's decision to slow down the U.S. breeder program rests on a conclusion that, in view of the associated nuclear weapon proliferation risks, a major effort should be made to delay as long as possible the advent of a plutonium economy, and on an assumption that, from an energy supply viewpoint, uranium resources will prove to be adequate to permit a substantial delay. Indeed, if alternative energy resources or alternative proliferation resistant nuclear fuel cycle technologies can be developed and deployed in time, it may be possible to by-pass the plutonium economy altogether.

The impact of U.S. deferral of its breeder program on foreign breeder reactor developments and commercialization schedules remains to be seen. Thus far, the preliminary reaction in other countries has been to press ahead with their own plutonium breeder programs. However, some or all of these national programs may yet encounter major technical difficulties which will cause slippage in current schedules, costs may increase so much that early commercialization will not be practical, or domestic opposition and international pressure against a plutonium economy may increase to the point where delay will occur because of political reasons.

In any event, the U.S. nuclear power research and development effort will be redirected towards the evaluation of alternative fuel cycles that are deemed to be more resistant to the risks of nuclear weapon proliferation.

There is only one possible way in which nuclear power might be developed which would avoid both the uranium resource constraint and a plutonium breeder reactor economy. This would involve a fundamental shift from the uranium–plutonium fuel cycle, on which the worldwide nuclear-power industry has been largely based, to the thorium-uranium-233 fuel cycle. The technical and economic feasibility of such a radical midcourse shift needs further study before it can be seriously considered. If the main reasons for wishing to escape from the plutonium economy are its diversion and routine handling implications, it is by no means clear that the situation would be improved by long-term reliance on a uranium-233 economy. Uranium-233 is a fissionable material which is about as useful for making nuclear explosives as plutonium, but it would be accompanied at all stages in the fuel cycle by uranium-232, an isotope which decays to form isotopes that emit highly penetrating gamma radiation.

The preceding paragraphs have shown that the way in which nuclear fission is most likely to become a long-term energy option is through the successful development of a plutonium breeder reactor. The prospect of a large, mature nuclear power industry based on plutonium raises profound issues for society aside from the energy that will be produced. Very large amounts—tens of thousands of kilograms—of plutonium will emerge annually from reprocessing plants, flow through fuel fabrication facilities, and be recycled into breeder reactors. Can a reactor which uses liquid sodium as coolant and fast neutrons for fission be made safe? Can toxic plutonium be managed in reprocessing and fuel fabrication operations without undue risk to the health of workers? Can large flows of plutonium—a few kilograms of which are sufficient for a nuclear explosive—be managed so that no significant amounts are stolen or diverted for use in nuclear weapons?

These issues are in addition to the long-term radioactive waste implications of a plutonium economy. All of them are embedded in the basic reprocessing decision. With these overarching considerations in mind, we shall now examine in greater detail the waste management implications flowing from the reprocessing decision.

No Reprocessing: The Once-Through Fuel Cycle

As mentioned earlier, when spent fuel is discharged from a power reactor it is stored at the reactor site in water-filled basins. The water acts as a heat transfer medium for the radioactive decay heat that must be removed from the fuel, and also provides shielding and a secondary containment barrier for any radioactivity that the fuel cladding fails to contain.

If the decision is made not to reprocess, the spent fuel cannot be left where it is for very long. Water basin storage of unpackaged, irradiated spent fuel is an unreliable method of isolating the contained radioactivity for periods longer than a few years, primarily because of deterioration in the quality of the containment provided by the spent fuel cladding. Economies of scale are likely to show that long-term storage of spent fuel should be carried out at one or more central locations rather than at each reactor.[1] Hence, after a period of cooling at a reactor site, spent fuel assemblies would be shipped in specially designed casks to these storage facilities. In any case, whether storage is centralized or dispersed, packaging will be required to provide an extra degree of containment. For spent LWR fuel, there is likely to be a period of three to four years between reactor discharge and packaging so that the provision for radiation shielding and heat removal in the package design is made easier.[2]

The question of whether to store the fuel retrievably or dispose of it irretrievably arises at this stage. The difficulties created by trying to establish a clear distinction between these states will shortly be discussed. Nevertheless, addressing the question at its face value for the moment, it could be argued that, in a society that is becoming acutely conscious of energy shortages, it would seem unlikely that an irretrievable disposal technology will be chosen for the spent fuel, despite the judgment that resulted in the initial decision not to reprocess. Irretrievable disposal would deprive future generations of the opportunity to resurrect a long-term nuclear option using the spent fuel inventory of fissile material. On the other hand, it could also be asserted that acute awareness of energy shortages does not presently appear to be effective in ensuring that future generations will have the opportunity to utilize oil and gas resources.

The technologies for irretrievable disposal of packaged spent fuel are similar to those for the disposal of HL waste in geologic formations. These will be discussed later in the chapter.

Various methods for retrievable storage have been proposed, including water-cooled basins, air-cooled vaults, concrete surface silos, near-surface heat sinks and geologic formations. Any method adopted must contain the radioactivity in the fuel and protect the fuel against mechanical, chemical, or thermal damage. It must also provide a safe subcritical arrangement for the fuel so that the risk of a fission chain reaction is acceptably low. This environment must be maintained under normal conditions and the facility must be designed to withstand a "maximum credible accident," such as the kind of natural catastrophe most likely in the area of the site.

An integral part of the design of a retrievable storage facility will obviously be the provision for retrievability. It is unclear as to how long it is feasible for a facility to retain this characteristic. A period of one hundred years is often mentioned as being practical for near-surface or surface engineered storage. However, retrievability is a feature that will not disappear overnight. Instead, inevitable deterioration of the various barriers will gradually make the recovery process more and more difficult. Therefore, in the case of a water-cooled basin, air-cooled vault, concrete surface silo or near-surface heat sink, the spent fuel would need to be removed before storage becomes "irretrievable." For geologic storage concepts, however, this may not be necessary. It should also be remembered that over the life of the storage facility the perception within society of what may or may not be retrievable is likely to vary with the levels of technological ability and the incentives for retrieval.

One result of reprocessing and recycling plutonium is that only about one-hundredth of the plutonium that would otherwise have to be stored in spent fuel is discharged in the HL and TRU waste. The plutonium in an unreprocessed, spent fuel assembly would take almost two hundred thousand years longer to decay to any given level than the plutonium discharged in the HL and TRU waste that would otherwise be generated by reprocessing that assembly and recycling the recovered plutonium. The implication sometimes drawn from this is that reprocessing irradiated fuel and subsequent fissioning of about 99 percent of the contained plutonium is advantageous from a long-term waste management perspective. This may be true, but the argument is not so simple.

For instance, irradiation in an LWR of a fuel assembly whose fissile enrichment is provided by recycled plutonium results in the production of more than ten times as much of the higher actinides (mainly americium and curium) as is produced by the equivalent irradiation of an assembly enriched in uranium-235.[3] Unless these isotopes—all of which either are themselves long-lived or decay to other long-lived species—are extracted from the waste and eliminated, part of the long-term management advantage disappears. Furthermore, it has already been noted that a decision to reprocess and recycle plutonium, combined with the development of a breeder, would allow nuclear fission to become a long-term energy option. Therefore, rather than making the very long-term comparison on the basis of a single assembly, it would perhaps be more reasonable to compare the implications of storing the spent fuel accumulated during a short-term nuclear option with those of storing HL and TRU waste generated during the development of a long-term nuclear power industry that is based on breeder reactors.

Reprocessing: The Closed Fuel Cycle

Having outlined some of the important issues arising from a decision not to reprocess, we now turn to the radioactive waste implications of the alternative. The discussion continues under the assumption that a decision to reprocess spent fuel is accompanied by a decision to recycle uranium and, sooner or later, plutonium. Unless otherwise noted, the discussion assumes plutonium recovery occurs during reprocessing.

From a waste management perspective, this route could be viewed as increasing the complexity of subsequent operations. Before reprocessing, almost all of the fission product and transuranic radionuclides are retained within the spent fuel assemblies by the cladding. Reprocessing involves chopping up the fuel, releasing fission product gases, leaching out the spent fuel core material from the cladding hulls with nitric acid, and subjecting the fuel solution to a series of chemical extractions designed to recover and purify uranium and plutonium. Each of these operations results in additional dissemination of the radionuclides originally contained in the spent fuel and the consequent generation of many waste streams with different physical forms and chemical compositions. However, it must be remembered that during reprocessing almost all of the by-product plutonium is extracted. As noted in Chapter 2, plutonium is extremely radiotoxic, has a half-life of about twenty-four thousand years, and is a major contributor to the long-term risk from post-fission wastes. After extraction it may be recycled and fissioned into shorter-lived fission products.

A brief review of the HL and TRU waste steams generated during reprocessing follows, with the emphasis on waste management.

After the chopping and leaching, the pieces of fuel cladding (hulls) remain undissolved. In addition, residual traces of undissolved spent fuel remain on the cladding hulls. Up to 0.1 percent of the fission products and actinides are trapped in this way.[4] The chop-and-leach operation results in the release of gaseous wastes.

Plutonium and uranium are extracted from the nitric acid fuel leach solution into an organic solvent. The aqueous solution remaining after this first cycle coextraction of uranium and plutonium is the principal component of the HL waste stream. The major constituents of HL waste are the fission products and the actinides. Over 99 percent of all the nonvolatile fission products contained in the spent fuel appear in the HL waste.[5] It is expected that approximately 0.5 percent of the uranium and plutonium contained in the spent fuel will not be recovered.[6] Thus, this fraction will appear mainly in the HL waste stream, together with almost all of the other actinides (the important ones being neptunium, americium, and curium).*

The properties that are important in the subsequent handling of the HL waste stream are the thermal power and the curies of radioactivity generated by decaying fission products and actinides, and the neutron emission rate from spontaneous fission. It should also be noted that the chemical form of the liquid HL waste stream is a nitric acid solution.

Finally, solid TRU waste in addition to the cladding hulls and assembly pieces emerges during the course of reprocessing. This waste may be divided into a number of different categories: combustible trash; noncombustible solids; slurries, sludges, and resins from water cleanup; and filters from gas cleanup. (It should be noted that low-activity aqueous waste, containing only trace quantities of transuranics, is also generated during reprocessing.)

At this point, a brief digression is appropriate. Modifications can be made to existing procedures or equipment that facilitate the management of wastes generated during reprocessing. These may not, strictly speaking, be considered as waste treatment technologies since, rather than actually handling waste streams, the waste management problem is affected in its gestation period. The major thrust of improvements currently being considered is to develop ''low chemical additive'' flow-

*An AEC report has stated that 99.5 percent recoveries of uranium and plutonium represent the practical limits of the reprocessing methods currently in use.[7] Statements such as this, combined with empirical evidence from the Nuclear Fuel Services reprocessing plant, which did not achieve these recovery efficiencies, lead some observers to feel that expectations of a 99.5 percent recovery are unduly optimistic.

sheets, and wherever possible to avoid unnecessary secondary waste streams.[8] The quantity and variety of wastes generated is thus minimized and the chemical composition aids subsequent operations.

This general principle—prevention rather than correction—may be extended beyond reprocessing to all sources of radioactive waste cycle. It is impossible to avoid completely the generation of waste in the fuel as long as there is fission, but much can be done to affect the type of waste that is generated.

Commercial HL Waste

Now let us focus on commercial HL waste. The structure of the following discussion corresponds to the decision scheme in Figure 2.1 (on page 25). It should be reemphasized that the order in which these decisions are actually made will not necessarily be the same as the order in which they are presented here.

Waste Partitioning

After reprocessing, the next strategic decision is whether or not to alter the chemical composition of the HL waste that emerges. Any operation which changes the waste composition will take place in the liquid phase. Therefore, although it may be necessary to store the liquid HL waste temporarily to allow further radioactive decay before such an operation takes place, it will almost certainly precede waste solidification, since performing further radioisotope extraction operations directly after reprocessing seems more sensible than first solidifying the waste, subsequently redissolving it, and then changing the waste composition. However, in the future a revised sequence may, for as yet unknown reasons, become attractive.

Both radioisotope extraction and solidification will occur at the reprocessing plant site since the transportation of liquid HL waste is prohibited for safety reasons. Such a prohibition is appropriate, since the risks of transport in liquid form outweigh the gains, if any, of waste treatment offsite.

Why should the HL waste composition be altered after it emerges from the reprocessing plant? Two different rationales are proposed.

First, the removal of certain radioisotopes from the waste may facilitate subsequent waste storage and disposal. Extraction for this purpose is generally referred to as "waste partitioning." Second, the waste contains isotopes that could be recovered and used in other applications. Extraction for this purpose may be called "isotope mining." Of course, waste partitioning and isotope mining are not mutually exclusive, since an isotope might, for example, be partitioned to reduce long-term risks from the waste and might also have commercial value.

With respect to partitioning, we observe that spent fuel reprocessing technology is governed by the law of diminishing returns. With available technology, it is not likely that it will pay to extract more than about 99.5 percent of the uranium and plutonium from the spent fuel. The extra cost of more complete extraction would more than offset the value of the energy in the fuel material recovered by the effort.[9] Therefore, as previously noted, the HL waste contains about 0.5 percent of the plutonium and uranium contained in the spent fuel as well as almost all of the fission products and transuranic actinides apart from plutonium. In this regard, it should also be pointed out that even if economic constraints did not apply and the sole reason for recovering plutonium were to reduce its contribution to the long-term risk in HL waste, it would still not be worthwhile to recover more than about 99.5 percent of the plutonium. This limit on recovery is due to the fact that the 0.5 percent remaining is eventually augmented by a similar amount that "grows" into the waste from the decay of the transplutonic actinides americium and curium. To increase the plutonium recovery efficiency while leaving the other actinides in the waste would, therefore, be illogical.

All isotopes of the actinide elements in HL waste either have long half-lives or eventually decay (perhaps via one or more intermediates) to other isotopes with long half-lives. "Long" here can be taken to mean thousands of years. On the other hand, almost all of the fission products have much shorter half-lives and take much less time to decay to innocuous levels. (Iodine-129 and technecium-99 are two of the exceptions, with half-lives of 17 million and 212 thousand years, respectively.) The rationale for partitioning is as follows: if HL waste can be divided into fractions according to half-life, the flexibility of subsequent waste storage and disposal operations will be increased, since it will then not be necessary to isolate all the waste from the biosphere for the length of time dictated by the toxic lifetime of the longest-lived radionuclide present in the unpartitioned stream. Therefore, if the actinides and perhaps the long-lived fission products can be partitioned from the HL waste, the residual fission product waste will probably need to be isolated from the biosphere for only about one thousand years. This is a long time in human history, but a relatively short time in geologic terms. It is short enough to be confident about the continuing integrity of a geologic formation in which the waste may be stored.

The long-lived fraction partitioned from HL waste might then be eliminated. Transmutation and extraterrestrial disposal have been proposed as elimination methods. The former would essentially convert long-lived radionuclides into shorter-lived ones, and the latter would provide an isolation method of cosmic proportions. Both alternatives are still in an early stage of development and are not likely to play more than a peripheral role in waste management planning in the near future.

Strontium-90 and cesium-137 have also been suggested as candidates for partitioning from HL waste, since these isotopes contribute about 90 percent of the activity in the spent fuel after a few years of cooling and continue to dominate for the first several hundred years. During this time they are the most important heat emitters, dictating the thermal design of the waste storage repository. However, removal would not materially reduce the potential hazard posed by these radioisotopes, since it is impractical to apply either nuclear transmutation or extraterrestrial disposal methods to them.

Methods to partition the long-lived radionuclides with sufficiently high extraction efficiencies have not yet been demonstrated, but the indications are that if they are feasible they will also be expensive. The same holds for the additional fuel cycle operations that would be necessary for the transmutation option and the space technology required for extraterrestrial disposal. Furthermore, these new steps would introduce new, short-term, operational risks and would also generate new secondary waste streams containing long-lived radioisotopes. Although the total quantity of these radioisotopes would be reduced, they would be more widely distributed in waste streams other than HL waste.

Assessment of the partitioning/transmutation and partitioning/extraterrestrial disposal options will be a difficult task, since it involves comparing the potential benefits of a reduction in risk that will not begin to take place for almost a thousand years with the increased costs and short-term risks of the technologies that must be implemented if this is to be achieved.

Turning to isotope mining of HL waste, the production of isotopes as by-products of the nuclear power program will far exceed the amounts required for applications in most cases. The management of the residual waste will be effectively unchanged. Furthermore, use of the radioactive isotopes that are mined defers, but does not eliminate, their contribution to the waste problem, since they must eventually be disposed of unless they have decayed to harmless levels during the period of their use. Indeed, waste management may be complicated further, since the radiotoxicity (and sometimes the chemical toxicity) of these isotopes requires their use to be tightly controlled.

Waste Solidification

Whether or not some isotopes are extracted after reprocessing, the liquid HL waste stream is a hazard if it is not handled and processed properly. Solidification immobilizes the waste and reduces its volume, thereby diminishing the potential hazard.

Current regulations require solidification to take place no later than five years after the liquid HL waste is generated at a commercial reprocessing plant. A number of different types of solidified product are proposed, each of which can be prepared by a number of different chemical processes. Despite this proliferation of means and ends, solidification technology is already quite well developed, and some of the processes have been extensively applied. Calcination has been used to solidify some U.S. military HL waste for over a decade. Calcined waste is, however, quite leachable. For this reason, it is probably unsuitable for long-term waste disposition. Methods to produce vitrified waste have therefore been developed. The final product of vitrification is a glassy matrix containing waste. It is highly leach resistant and is regarded as being well suited for long-term isolation of waste. The technology for other solidified waste forms, for example, supercalcine, metal matrix, glass ceramic, and coated pellet, is at an earlier stage of development.

It has been suggested that, instead of solidification, HL waste might be disposed of simply by injecting it in liquid form into suitable rock formations. The decay heat from the waste would be sufficient to melt the surrounding rock. As the waste continued to decay, the molten rock-waste would eventually solidify, thus immobilizing the waste in a stable, nonleachable form. This *in situ* melting concept for liquid HL waste disposal could only be utilized if the reprocessing plant were sited above a suitable geologic formation, unless the present regulations are changed to permit transport of liquid HL waste. The concept is not generally favored, perhaps because it is so unequivocally irretrievable from the outset. It is worth noting at this stage that the *in situ* melting concept has also been proposed for the disposal of solid HL waste. Similar considerations apply, except that in this case collocation of the reprocessing plant and disposal site would not be required.

Interim Retrievable Storage

The next major decision concerns retrievable storage of solid HL waste in surface or near-surface storage facilities. After the waste is solidified and suitably packaged, it can be transported either to a permanent storage or disposal site directly, or first to an interim storage facility and then, at some later date, to a final resting place. Interim retrievable storage concepts place the responsibility for waste isolation on man-made systems maintained under appropriate surveillance. As noted previously for storage of unreprocessed spent fuel, ''interim'' generally is used to mean a period of up to one hundred years.

The idea of interim retrievable storage of HL waste has had a checkered history. The present situation is that an earlier decision to construct an interim storage facility for the waste has been reversed, and it is now planned to transport solidified HL waste directly to a permanent storage/disposal site. However, the retrievable storage concept may still be revived. Water-cooled basins, air-cooled vaults, and concrete surface silos have all been proposed as storage methods.

Interim retrievable storage of HL waste allows time for the development of new, improved permanent disposition technologies. Actinide partitioning–elimination schemes could also be developed during a period of interim storage. If the stored wastes are in solid form, however, subsequent redissolution for the purposes of partitioning could prove to be either technically unfeasible or too costly.

It is argued by some that, apart from the development of partitioning which would reduce the time required for waste isolation, scientific and technological progress that will take place during a period of interim storage will not reduce significantly the long-term risk of permanent and probably irretrievable disposition in geologic formations. If this is true then the incentives for a period of retrievable storage are greatly reduced. The same conclusion can be drawn by those who argue that there is already enough technical evidence to show that the risk presented by isolation in geologic formations is acceptably low. Furthermore, it is argued that even if significant progress is made during the interim period, society at that point may be unable to find the resources to retrieve the waste and store it permanently, if this is then regarded as desirable. The prospect of societal collapse is also invoked in this context. There is no reason to expect such social disintegration to wait until the permanent disposition of radioactive waste has been completed.

Others feel that a decision now to select and implement a final disposition method—a decision based on the supposition that current knowledge is adequate and unlikely to be improved significantly in the future—must be avoided. They argue that, despite the additional responsibility that is involved, future generations would prefer to be in a position to make waste management choices of their own, rather than to witness passively the irreversible consequences of a previous decision. But, of course, the implied assumption here is that our current assessment of what constitutes "irretrievability" will be shared by future generations, whereas it may be that recovery of waste from what would today be considered an irretrievable state might not present insurmountable obstacles for the increased technological capabilities of future generations, if such a task were then perceived to be necessary. Nevertheless, the argument is sometimes extended to propose the viability of a "permanent" retrievable storage option. Because of the limited time for which a single storage facility can retain its retrievability property, permanent retrievability can only be retained by continual renewal and replacement of the systems used to isolate the waste—a burden that is justified, in the minds of the proponents, by the fact that in a sense all options continue to remain open in successive generations.

The trade-off between the essentially irretrievable option of permanent geologic disposition and retrievable storage as a permanent strategy centers on the question of where reliance should be placed: on our existing ability to predict the evolution of geologic history over the next million years or so, or on the ability and motivation of future generations to manage the waste safely. From a different perspective one might argue that, in the case of permanent retrievable storage, primary reliance is placed on the present *and continuing* ability to predict the evolution of human history over much shorter periods. In this case, because of the periodic renewal of storage facilities that would be necessary, the time scales for prediction are measured in hundreds rather than thousands of years.

The distinction between immediate and essentially irretrievable geologic disposition and a period of (or perhaps permanent) retrievable storage is not as clear-cut as has been presented above. For instance, an intermediate concept would involve emplacing the waste in an engineered retrievable storage facility constructed within a deep geologic formation. The advantages claimed previously for retrievability would be retained, and, in addition, the geologic formation would provide a safety barrier over and above the man-made, primary containment. The man-made containment would be necessary for the retrievability provision, and it might perhaps be expected to survive for up to a hundred years or considerably longer if maintained and periodically renewed. The geologic formation could provide long-term containment if, at any stage, society were unable to continue to maintain the facility. Besides, the concept would be less susceptible to the disruption that might be caused by terrorist or military attack that is claimed to be a significant disadvantage of surface or near-surface storage facilities.

On the other hand, the cost of constructing and maintaining such an interim retrievable facility would certainly be greater than either the cost of a permanent geologic repository with no provision for retrievability or the cost of a retrievable storage facility located at or near the surface; and the risks to operating personnel might be greater than for either of the other two concepts. Furthermore, in spite of the extra degree of containment that would be provided in such a facility, the concern that has been expressed as to the adequacy of the sealing process in geologic repository concepts would be aggravated, since, by definition, if the waste were to be retrievable, access would have to be retained.

Permanent Disposition of HL Waste

Many persons with diverse backgrounds believe that the most important decision to be made in developing a radioactive waste management system is the choice of a permanent disposition method. It is easy to understand why: the waste must be isolated from man and the environment for as long as it will present a potential radiological hazard.

Unfortunately, one characteristic of choosing a permanent disposal method is that we can never know whether we have chosen safely, since such a long period of time is likely to elapse before errors in judgment or mistakes in implementation show up. However, this feature of decision making regarding permanent waste disposition appears more extraordinary than it really is. Most major social decisions are made under large uncertainty and create different distributions of benefits, costs, and risks within a society geographically and/or economically, if not across long stretches of time.

Here, we are concerned with what may be done to minimize the risk that something actually will go wrong.

A technical consensus appears to have been reached to the effect that the best place to isolate HL waste is under the earth's surface. The major debate concerns which areas of the earth's surface and which underground geologic formations are the most suitable. Continental land masses, the ocean floor, and polar ice caps are proposed as suitable areas. Since these three areas together include most of the surface of the earth, we must be more specific.

On land, a number of different geologic formations are being considered. Rock salt has been a candidate for over two decades. HL waste disposal in salt formations continues to be the favored alternative within the U.S. government. Hard, insoluble crystalline rock formations such as granite and basalt are in some ways preferable to plastically flowing, water-soluble salt, but in other respects they are less desirable. The same applies to formations of shale or sandstone.

More than two-thirds of the earth's surface is covered by the oceans. Some areas of the ocean's deep seabed are geologically and seismically stable, are remote, and are biologically relatively unproductive. These areas are, moreover, separated from man's immediate environment by huge volumes of a diluting medium. Thus, the deep seabed contains possible sites for HL waste repositories. The disposal concept is not ocean dumping, but rather controlled emplacement of solidified, packaged, HL waste in either the argillaceous sediment or the underlying basaltic bedrock. The engineering capability to perform the required ocean floor operations is quite highly developed. Therefore, seabed disposal appears technically feasible. However, more knowledge is needed concerning the physical, chemical, and biological mechanisms by which radionuclides in HL waste could be released from the repository and transported through the seabed and into the marine environment.

HL waste storage or disposal in or under the Antarctic ice sheet is an alternative that is currently not considered as attractive as deep continental or seabed disposal. Most of the schemes envisioned would allow the waste to melt its way through the ice sheet, eventually coming to rest either at the ice–rock interface or within the bedrock itself. Other concepts include the construction of an interim surface storage facility which, after decommissioning, would become covered with snow and would subsequently sink through the ice toward the bedrock.

The previous discussion has focused on the technological implications of the decision sequences involved in the implementation of various commercial HL waste management strategies. Some of the more important risks, costs, and benefits arising from these decisions have been identified. An in-depth comparison, based on detailed and comprehensive studies, will be necessary to provide the foundation for choosing which particular strategy or strategies should be implemented.

Military HL Waste

This section discusses the military HL waste situation and various ways in which it might be handled in the future. First, we discuss what may be done with the waste that has already accumulated. Then we consider the alternatives available for managing military HL waste generated in the future. The majority of the future waste will come from the continuing plutonium production program. The division is somewhat artificial, since management policy for waste in prospect is heavily influenced by previous practices. Nevertheless, by structuring the discussion this way, it is easier to focus on the important waste management issues. Figure 2.2 (on page 00) shows diagrammatically the management strategies currently applicable for military HL waste.

Existing Waste Treatment

In Chapter 1, the quantities of existing military HL waste were briefly outlined. This information will now be augmented.

By the early 1980s, it is estimated that approximately 65 million gallons of HL waste will exist in solid form. An additional 9 million gallons will be "in process" as liquid that is being allowed to decay before solidification and liquid that cannot be solidified with present technology.[10] At the present time, 72 percent of the military HL waste is stored at Hanford, 25 percent at Savannah River, and 3 percent at Idaho Falls.[11] These ratios will not change appreciably by the early 1980s.

At Savannah River and Hanford, with the exception of the HL waste produced in the very early reprocessing plants that have long since been shut down, almost all HL waste is originally generated in acid form. Acid liquid HL waste must be stored in stainless steel tanks. In the immediate post-World War II years stainless steel was expensive and not easy to obtain. The decision was therefore made to neutralize the wastes. This substantially increases the liquid waste volume and causes most of the more radioactive isotopes (with the exception of cesium-137) to precipitate out in a sludge. Nevertheless, neutralized waste can be stored in carbon steel tanks. These were (and still are) cheaper than stainless steel tanks, and were therefore used for storing the sludge and liquid after generation. The lifetime of carbon steel tanks was shorter than expected and leaks occurred.

One possible way to deal with the leaks would have been to reduce the working lifetime of the tanks; this could have been done by building new ones and periodically transferring waste from one tank to another. An alternative procedure was to solidify the waste, thus stopping the leaks and, since solidification reduces volume, creating extra storage space in existing tanks. The latter alternative was cheaper, and it was selected.

Solidification of neutralized liquid HL waste is achieved by evaporating it down to a damp salt cake. Evaporation does not solidify all of the liquid, however, and there is always a quantity of residual liquor remaining. (Before solidifying the Hanford waste it is necessary to partition most of the principal heat-emitting isotopes, strontium-90 and cesium-137, since the tanks in which the solidified waste is stored, unlike those at Savannah River, are not equipped with cooling coils. These isotopes are encapsulated in solid form and currently stored in water basins.) The solidification program will not reach a steady state until the early 1980s, by which time there will be nearly 500,000 tons of residual solids in the tanks.[12]

What should be done with these solids? Should they remain where they are or should they be removed?

One possible course of action is to leave them where they are without further treatment. It is quite conceivable that such a plan would be deemed unacceptable. Salt cake is extremely soluble, the carbon steel tanks are corroding, and the arrangement would provide reliable containment for no more than a few decades. Another alternative is to leave the solids in the tanks and treat them further. The salt cake might be converted into a form less susceptible to leaching. Then the tanks might be covered with asphalt or concrete to make them more inaccessible to water and living organisms.

If the solids are to be removed from the tanks, the problem of extraction must be faced. Two methods are currently being considered. Neither seems very attractive. High-pressure water jets might be used to sluice out the partially dissolved solids. Such a procedure would not be suitable for the many tanks (particularly at Hanford) that are cracked because dissolution of the salt would reopen the previously blocked cracks and cause still more leakage. Alternatively, the solid salt cake might be dug out of the tanks—a risky procedure made more difficult because access to the tanks is currently limited to a few narrow vent pipes. Furthermore, the excavation work would increase the risk of releasing airborne radioactivity.

After it is removed, what should be done with the salt cake? It will probably require further treatment to produce a solid form that is more suitable for permanent disposition. The technology required for this step is not yet developed. It is not clear whether there is a form that fulfills the dual requirements of long-term stability and the ability to be produced from salt cake.

Finally, the costs of treatment and removal of existing military HL waste from tanks at Hanford and Savannah River must be mentioned. Cost estimates vary over a wide range from 2 to 20 billion dollars just to prepare the existing military HL waste inventory at Hanford and Savannah River for permanent disposition. Two conclusions seem appropriate: (1) the cost of managing military HL waste is likely to be greater than the cost of commercial waste management for quite some time to come; (2) in radioactive waste management, the longer-term consequences of decisions made only on the basis of short-term considerations can be enormous. It should also be remembered that "longer term" here means about fifty years—a very small fraction of the toxic lifetime of the waste.

With regard to existing solid HL waste, the situation at Idaho Falls is significantly different from Hanford and Savannah River. At this site, the acid waste has not been neutralized. Instead, it has been stored temporarily in stainless steel tanks, from which there have been no leaks. Since the opening of the waste calcination facility in 1963, it has been processed to form a dry granular solid called waste calcine. The calcined waste is stored in stainless steel bins partially underground. The lifetime of these bins is estimated to be several hundred years.

Calcined waste, however, is relatively leachable. Bin storage cannot therefore be relied upon to provide adequate isolation for the toxic lifetime of the waste. Thus the question of when to begin removal will have to be settled. Unlike the situation at Hanford and Savannah River, the removal of solid wastes from storage bins will not be a very difficult technical operation. In view of the technical similarity between the waste produced at Idaho Falls and that from the commercial power reactor program (they are both retained in acid form and then calcined), the waste management decisions made for each should presumably be consistent. For example, it would not seem consistent to decide upon an extended period of storage for calcined waste in bins at Idaho Falls and also decide that retrievable engineered storage is not suitable for commercial HL waste.

Management of Future Wastes: Short-Term

Major modifications to existing management practices for HL waste at Idaho Falls do not appear necessary in the immediate future. The important decisions yet to be taken concern the post-calcination steps and apply equally to existing and future waste.

This is not the case at Hanford and Savannah River. The root of the present problem—and its magnitude is considerable—seems to have been the initial decision to neutralize the acid waste. This practice continues. The issue is: should future acid HL waste be neutralized? At this stage, it is helpful to briefly compare a neutralized-waste/salt-cake system with an acid-waste/calcine system.

As discussed previously, the initial decision to neutralize was based on the availability and cost of carbon steel compared to stainless steel. The former is still cheaper than the latter, but it should also be remembered that neutralization results in a significant increase in volume, so that the tankage required for acid storage is less than for storage as a neutralized solution. In addition, the lifetime of stainless steel tanks is longer than that of carbon steel tanks used for waste storage purposes.

Solidification technology is different for the two systems. Calcined waste, although not suitable for long-term storage because of its leachability, is a satisfactory base from which to produce glasses and other more suitable forms. Calcination of acid waste has taken place routinely since 1963. On the other hand, it is not possible to calcine neutralized wastes with present technology, and the evaporation/crystallization method used to solidify waste at Savannah River and Hanford is not entirely satisfactory. In addition, salt cake is more soluble than calcine and is even less suitable for long-term storage. A method for converting it into a less leachable form such as glass has not yet been demonstrated on a commercial scale, and it will be a difficult problem to solve. The volume of salt cake per unit of spent fuel reprocessed is significantly greater than the corresponding volume of calcine produced by an identical amount of spent fuel. Finally, calcined waste in bins at Idaho is much more retrievable than salt cake, especially the material in the flawed tanks at Hanford.

With this comparison in mind, we may consider the question of whether or not to continue to neutralize future HL waste at Hanford and Savannah River. At Hanford, the military production of plutonium is gradually being phased out. Of the nine production reactors that have operated at one time or another over the past thirty years, only one is still functioning, and that one will possibly be shut down within a few years. Therefore, in view of the relatively small quantity of HL waste that is still to be generated at Hanford, a switch to an acid waste management system at this stage may be uneconomical. The Savannah River plant is now the major plutonium production facility in the U.S., and it will continue to generate HL waste in relatively large quantities for as long as plutonium is required for nuclear weapons. In this case, it may be preferable to convert to an acid/calcine system. Nevertheless, a recent assessment, based upon economic estimates to the year 2000, indicates that it will be cheaper to continue with an improved neutralization/salt-cake system than it would be to convert to an acid/calcine system.[13]

Permanent Disposition: Existing and Future Waste

It is possible that existing military HL waste, in salt-cake form, will be left in existing tanks more or less permanently. No decision as to the most suitable long-term strategy for the waste has yet been made. Indeed, the period of formal deliberation that will precede this decision, beginning with a presentation of the available technical alternatives, is scheduled to be initiated in 1977. Considerable delay is possible, perhaps likely, in view of the costs involved.

After HL solid waste has been removed from the tanks and converted into a different form, it is necessary to decide on a site for final disposition and then to package and transport the waste to the site chosen. (Retrievable interim storage is also a possibility, but it would be an additional, costly stage in an already very costly program, and, depending on the performance of the existing temporary storage facilities over the next few decades, it might also be unnecessary.) The location could be either on or off the tank storage site. At the moment, on-site disposition seems unlikely. Earlier AEC efforts to establish the feasibility of on-site geologic disposal of the solidified HL waste at both Hanford and Savannah River were suspended some time ago. (Recent reports indicate, however, that investigations at Hanford have resumed.[14]) The alternative off-site permanent disposition methods for solidified military HL waste are similar to those for solidified commercial waste. However, alternatives that are international in scope, specifically deep seabed and ice-sheet disposal, may not be available for military waste disposal if other governments raise political and legal objections.

Demonstration of Permanent HL Waste Repository

ERDA has recently initiated its National Waste Terminal Storage Program, in which the current research and development phase of radioactive waste disposal activities will evolve into a system capable of dealing with the final disposition of HL waste from the nuclear power industry.[15] An integral part of this program will be the pilot-scale demonstration of a number of HL waste repositories.

The target date for the initial waste repository demonstration to start is 1985. It will begin with the establishment of a pilot facility, to be followed by expansion and then, if nothing unforeseen has occurred and the demonstration has been successful, conversion into a full-scale federal repository. The change from an expanded pilot plant to a full-scale federal repository will be characterized by the foreclosure of the retrievability option. Storage rooms will be backfilled and sealed. Six sites for pilot repositories are to be established on a staggered basis, with the first two in salt and ready to receive waste in 1985, and subsequent installations in formations other than salt. The impact of President Carter's new nuclear policy on these plans for demonstration projects is unclear.

What are the technical benefits from operation of a pilot-scale plant before it is converted into a permanent repository? Experience will be gained in the handling of packaged solid waste; in the construction and operation of surface and underground facilities for receiving waste; in the emplacement of waste in the geologic formation; and in the measurement of the physical effects of waste emplacement caused by heat and radiation fields and mining stresses. It will, of course, be impossible to "demonstrate," with high assurance of validity, the capability of the repository to contain HL waste over the period for which it constitutes a potential radiological hazard. What can be demonstrated is the ability to receive and emplace solid waste in the repository. In the technological dimension, confidence in the long-term reliability is more likely to stem from the geologic and hydrologic measurements preceding repository site selection. Such measurements are intended to provide reliable information as to the age and stability of the formation in question.

What are the possible nontechnical benefits from the demonstration? It seems generally acknowledged that one of the major obstacles to public acceptance of nuclear power is concern that there is no safe solution to the radioactive waste problem. It is believed by many that this concern is derived, at least in part, from the absence of a demonstrated permanent waste repository. It seems reasonable to conclude that one of the most important purposes intended by the waste repository demonstration "on a timely basis" (i.e., in operation by 1985) is to increase public confidence.

But what worries the public about the disposal of radioactive waste? It is the concern that, in pursuit of our objectives of providing ourselves with energy and security, we might also be responsible for harm inflicted on the next generation or more far into the future by failing to provide adequate isolation of this radiologically hazardous material.

Assuming the demonstration is successful within its limitations, what will it help to achieve in public confidence? It is possible that a demonstration will satisfy the public that the task of radioactive waste management can be dealt with safely. But if this occurs, the demonstration will have created an "illusion of certainty" (to quote Kenneth Boulding's phrase). Even those who have great confidence in the ability of our society to dispose of radioactive waste in a safe manner would not claim that the operation of a waste repository for a few years will, in itself, provide substantial evidence to support a conclusion that the waste is very unlikely to harm man or the environment thousands of years in the future. Should we proceed on the basis of such an illusion of certainty in the public mind, even if the illusion is not shared by those in authority?

Another possibility is that the general public will become more sophisticated in its ability to make judgments on the basis of technical and probabilistic information, and the predominant view will be that the risks of radioactive waste disposal are acceptable. If such a development actually takes place, the public will be unlikely to regard the demonstration as being of the utmost significance as an input into a long-term decision-making process. A third possibility is that there will be a public reaction against what is seen as an attempt to manipulate opinion with the demonstration. In this event, the obstacle which the radioactive waste problem poses to public acceptance of nuclear power will become even greater. It may be concluded from this discussion that it would be unwise to rely too heavily on the successful demonstration of pilot permanent waste repositories to obtain public acceptance of the waste management program.

Finally, we consider a related but broader question: what is the optimum rate of implementing waste management strategies? Technological constraints, such as the time required for research and development, and physical requirements for waste handling or lack thereof, enter into the decision. Delays in commercial reprocessing would permit delays in implementing commercial HL waste management strategies. But the perceptions of the general public as to the urgency of demonstrating solutions to radioactive waste problems are also important, and these do not necessarily correspond to the technical demands of the situation. Increasing the rate of implementation may be perceived to be unsatisfactory, even if it is technologically quite feasible. On the other hand, the public may begin to equate "no decision" with "no solution," and, to the extent that the tendency has political significance, its effect will be to accelerate implementation, leading to the possibility of premature resolution of issues from a technological viewpoint.

Current public perceptions are complicated by a history of institutional torpor in which the AEC, influenced by the slow development of commercial reprocessing, pursued a slowly paced waste management program while perhaps failing to recognize the importance of developing an approach that was generally acceptable to the public, even before it became an immediate technological requirement. The argument that nuclear power should not be used to generate electricity until a safe method for long-term radioactive waste disposal had been demonstrated seems to have been largely unanticipated. This is perhaps understandable in view of the lack of substantial public concern about military waste management operations during three decades of the buildup of nuclear armaments.

In any event, ERDA's terminal storage plan, involving the development of multiple sites and broad public participation in the major decisions, now appears calculated to balance technological and political factors in demonstrating that there are solutions to the long-term radioactive waste management problem.

Commercial TRU Waste

Under existing ERDA standards[16] and proposed NRC regulations,[17] TRU waste is defined as low-level waste which is presumed to contain more than ten nanocuries of long-lived transuranic radioactivity per gram. This definitional limit is currently being studied and may be revised upward or, possibly, completely redefined.

Low-level, transuranic contaminated waste (TRU waste) will require more careful management than other kinds of low-level waste because of the very long time for which it will remain a radiological hazard. Indeed, assuming that reprocessing and recycling using existing process methods are introduced, more than half of the plutonium released in nuclear fuel cycle wastes will be contained in low-level waste streams.[18] However, although TRU waste contains at least as much plutonium as HL waste, the plutonium concentration is very much lower. Nevertheless, in basic policy decisions involving long-term storage criteria for TRU waste, the quantity of long-lived radionuclides may be given as much emphasis as the concentration.

TRU waste streams of three types—noncombustible solid, combustible solid, and liquid—are generated by three stages of the commercial fuel cycle—reprocessing, the conversion of liquid plutonium nitrate to solid plutonium oxide, and mixed-oxide fuel fabrication. Measured by the amount of plutonium contained in TRU waste per unit of electric power produced, TRU waste is created in these three stages at roughly the same rate.[19] In similar units, each type of TRU waste—combustible and noncombustible solids and liquid—is produced at approximately the same rate. These two facts imply that all TRU waste streams from all TRU waste-producing stages of the fuel cycle should receive roughly equal attention.

With this brief introduction in mind, we will discuss the major TRU waste management decisions. As with HL waste management, all decisions are interdependent.

The management of TRU waste can be affected by operational changes at the points of generation. As in the case of HL waste management, this concept is one of prevention rather than correction. For instance, the efficiencies of plutonium recovery at the reprocessing plant* and plutonium utilization at the mixed oxide fuel fabrication plant could both be improved, so that less plutonium is lost in the TRU waste streams. The argument used earlier for plutonium recovery from HL waste involving the law of diminishing returns applies here. It may be economic to remove part of the plutonium contained in some TRU waste and recycle it to reactors. But in order to obtain a significant reduction in the long-term environmental risks from these wastes, it will be necessary to achieve a higher recovery efficiency. The cost of these operations will not match the value of the extra fuel obtained as a result. It therefore seems likely that the reprocessor and fabricator would require some form of encouragement—a government subsidy as a carrot or regulatory standard as a stick—if these improvements are to be achieved. Similar considerations apply to the flowsheet modifications that could be made at reprocessing and mixed oxide fuel fabrication plants to make the physical form and chemical composition of the TRU wastes more suitable for subsequent waste management operations.

The issue of improved plutonium recovery is related to the HL waste-partitioning question. The ability of waste partitioning to reduce the long-term risk of HL waste management has already been discussed.

*For convenience, the nitrate–oxide conversion facility will now be considered as part of the reprocessing plant. In practice, the two functions, reprocessing and conversion, will essentially be combined into one process flowsheet.

Since the amounts of long-lived radionuclides in unpartitioned HL waste and untreated TRU waste are similar for each unit of electricity produced, it could be argued that a decision taken to partition HL waste, if consistent, would have to be accompanied by a decision to remove enough of the transuranics (mostly plutonium) from the TRU waste to achieve a corresponding reduction in long-term risks.

Transuranic recovery from TRU waste would precede or follow a number of other treatment steps. For instance, sorting and shredding followed by incineration of combustible waste produces a residue (or ash) that can be processed for plutonium recovery. An incentive for incineration (or some other method of combustion) is the volume reduction that is achieved. Reduction factors varying between 20 and 50 are claimed for the different processes. Mechanical compaction can also be used to reduce the as-generated volume of both combustible and non-combustible solids.

In view of the great variation in TRU waste streams generated in reprocessing and mixed oxide fuel fabrication plants, it is not surprising that there are a great many different treatment technologies. Some of them are still at an early stage of development, probably because the commercial nuclear industry has not begun to generate TRU waste in significant quantities, and until recently TRU and non-TRU low-level wastes were not distinguished in management operations involving military and research and development waste. Others are equally applicable to TRU and non-TRU wastes, and certain of these have been used routinely for some time. A substantial proportion of these generally applicable non-high-level waste treatment technologies has not yet been introduced into commercial waste management systems.

Nevertheless, all of these technologies are designed to achieve one or more of the following objectives: the creation of a waste form with good long-term storage properties; volume reduction; the reduction of short-term risks. In all cases, the importance of minimizing secondary waste stream production is clear, since each operation involving the handling of radioactive materials tends to disseminate the radionuclides in the feed stream among a number of different output streams.

The final two stages in TRU waste management—each of which bears directly on the long-term risk posed by this waste—are immobilization and final disposition. They must combine to prevent the migration of radionuclides from TRU waste to the biosphere. Again, in view of the wide variety of TRU waste forms, there will be a number of applicable immobilization technologies.

It is becoming increasingly likely that the long-term isolation criteria that will be applied to HL waste will also be applied to TRU waste. The logic behind this trend is based on the realization that the amounts of long-lived radioactivity in these two waste categories are comparable, although the concentrations are, of course, significantly different. Decisions concerning the permanent disposition of HL and TRU wastes will be closely related, and the options presented in the section on HL waste are applicable.

Military TRU Waste

Military TRU waste comes principally from two sources—plutonium separation from irradiated plutonium production reactor fuel and the fabrication of nuclear weapons. Information concerning the latter is classified. As has already been mentioned, until 1970 TRU waste under the jurisdiction of the old AEC (most of which had a military origin) was emplaced in shallow burial grounds with no provision for retrieval. Since then, TRU waste at ERDA (formerly AEC) burial sites has been stored retrievably above ground.

Taking the long-term perspective, particularly with regard to permanent disposition, it seems unjustifiable to manage the military and commercial TRU wastes so that the risks posed by one will be greater than those posed by the other. The waste must be managed consistently, regardless of origin. In the case of the weapons fabrication waste this may present security-related problems. But these difficulties can be solved administratively, and they should not obscure the fundamental waste management issue—the long-term nature of the risk.

Conclusions

From the preceding discussion of radioactive waste management strategies, the following conclusions emerge.

1. The decision whether or not to reprocess commercial power reactor fuel has fundamental importance for the future development of nuclear fission as a source of energy. Nuclear fission is likely to be a long-term energy option only with reprocessing *and* successful breeder

reactor development. Either the uranium/plutonium, or the thorium/uranium-233 fuel cycle, or a combination of both, is possible as a long-term fission energy option, but the uranium/plutonium fuel cycle is the most likely to mature first. With reprocessing and plutonium recycling in existing commercial LWR's only, the relatively short lifetime of the fission energy option based on the once-through fuel cycle is only marginally extended.

2. Radioactive waste management is an important, though not in itself decisive, issue in the reprocessing decision. If commercial spent fuel is not reprocessed, over the long term it will be a major waste management problem. Some form of reprocessing may eventually be necessary as a waste treatment measure, if not to recover uranium and plutonium for use as fuel. If commercial spent fuel is reprocessed for the purposes of fuel recycling, then HL and TRU waste streams, which emerge principally during reprocessing and uranium/plutonium mixed oxide fuel fabrication, will be interrelated long-term waste management problems.

3. The decision whether or not to reprocess military fuel was made at the outset of the U.S. nuclear weapons program in order to obtain plutonium for use in weapons. From past military plutonium production, the legacy of liquid and solid accumulations of HL waste in temporary tanks constitutes a multibillion dollar treatment problem before the waste can be safely placed in a permanent repository.

4. Radioactive waste management operations are interdependent. The risks and costs of permanent disposition methods affect and are affected by particular waste solidification, packaging, and other treatment methods. In view of the long-term risks, essentially all HL and TRU waste, regardless of its commercial or military origin, should be eventually destined for one set of government-managed radioactive waste repositories.

5. Decisions regarding every phase of radioactive waste management must be made under uncertainty. Research and development can narrow technological uncertainty. Short-term, small-scale demonstrations of permanent waste repositories can show that radioactive waste can be safely emplaced in geologic formations, but little more.

6. Radioactive waste management requires comprehensive strategic planning. A key issue in such planning is the distribution of risks and costs through time. The balancing of multiple conflicting objectives may contribute even more to the difficulty of waste management deci-

sion making than the more familiar characteristic of technological uncertainty. In radioactive waste management, quality of thought is more important than timeliness in decisions, because the current costs of delay are likely to be much less than the future costs of error. This reasoning runs counter to the strong pressure which currently exists for an early "solution" to the radioactive waste problem.

7. From a technological standpoint, radioactive waste operations require a fully integrated management framework. The management framework should be vertically integrated so as to include all operations from temporary waste storage, through treatment and transport to permanent disposition. The framework should be horizontally integrated to include post-fission HL and TRU wastes of both commercial and military origin.

Radioactive waste management will be a continuing challenge for government throughout the nuclear age and perhaps for many centuries beyond.

Notes to Chapter 2

1. King, F. D. and Baker, W. H., "Interim Storage of Spent Fuel Assemblies," in *Proceedings of the International Symposium on the Management of Wastes from the LWR Fuel Cycle,* Denver, Colo., July 11–16, 1976, Conf-76-0701, U.S. Energy Research and Development Administration, Washington, D.C., 1976, 210.
2. *Ibid.*
3. Bell, M. J., *Heavy Element Composition of Spent Power Reactor Fuels,* ORNL-TM-2897, Oak Ridge National Laboratory, Oak Ridge, Tenn., May 1970.
4. Steindler, M. J. and Trevorrow, L. E., "Description of the Fuel Cycle and Nature of the Wastes," in *Proceedings of the International Symposium on the Management of Wastes from the LWR Fuel Cycle,* 87.
5. *Ibid.*
6. *Ibid.*
7. U.S. Atomic Energy Commission, *The Safety of Nuclear Power Reactors (Light Water Cooled) and Related Facilities,* WASH-1250, July 1973, 4–73.
8. Bond, W. D. and Leuze, R. E., *Feasibility Studies of the Partitioning of*

Commercial High-Level Wastes Generated in Spent Nuclear Fuel Reprocessing, ORNL-5012, Oak Ridge National Laboratory, Oak Ridge, Tenn., January 1975.

9. See footnote 7.

10. U.S. General Accounting Office, *Isolating High-Level Radioactive Waste from the Environment: Achievements, Problems, and Uncertainties,* Report to the Congress, B-164052, Washington, D.C., December 18, 1974, 20.

11. Dance, K. D., *High-Level Radioactive Waste Management: Past Experience, Future Risks, and Present Decisions,* Report to the Resources and Environmental Division of the Ford Foundation, April 1, 1975, p. 27.

12. See footnote 14, Chapter 1.

13. Crandall, J. L. and Porter, C. A., *Economic Comparisons of Acid and Alkaline Waste Systems at Savannah River Plant,* DPST-74-95-37, DuPont de Nemours, Aiken, S.C., 1974.

14. Kenneth R. Chapman, Director, Office of Nuclear Material Safety and Safeguards, U.S. NRC, personal communication, October 7, 1976.

15. An outline of ERDA's program plan can be found in: Zerby, C. D. and McClain, W. C., "Waste Isolation in Geologic Formations in the United States," in *Proceedings of the International Symposium on the Management of Wastes from the LWR Fuel Cycle,* 567.

16. U.S. General Accounting Office, *Improvements Needed in the Land Disposal of Radioactive Wastes—A Problem of Centuries,* Report to the Congress, B-164105, January 1976, 5.

17. 39 *Fed. Reg.* 32921 (1974).

18. Bray, G. R., "Other-Than-High-Level Waste", in *Proceedings of Nuclear Regulatory Commission Workshop on the Management of Radioactive Waste: Waste Partitioning as an Alternative,* Battelle-Seattle Research Center, Seattle, Wash., June 8–10, 1976, NR-CONF-001, USNRC, Washington, D.C., 1976, 102.

19. *Ibid.*

Chapter 3

Existing Policy, Law, and Organization

HAVING CONSIDERED THE CHARACTER of radioactive waste in Chapter 1 and the range of strategic decisions involved in waste management in Chapter 2, we are now in a position to analyze in some detail the existing situation in the United States regarding radioactive waste management and regulation. First, the policy goals are briefly discussed. Next, the organizational structure for management, regulation, and research and development related to radioactive waste is analyzed. Finally, the existing situation regarding implementation of radioactive waste policy is described.

Goals

Present statements of the basic policy goals of radioactive waste management vary somewhat according to the federal government agency concerned. The statements of the Environmental Protection Agency (EPA) and ERDA to date have focused mainly on technological safety. EPA's position is:

> Waste management means containment of radioactive materials until they have decayed to innocuous levels. The objectives that EPA has are to

minimize exposure to present and future populations and to avoid dilution into the biosphere.[1]

ERDA's is similar:

> . . .the effective management of nuclear wastes in a manner which effectively protects man and his environment. . .[2]

With respect to commercial HL waste in particular, ERDA's objective is:

> . . .to provide multiple terminal storage sites on time to receive solidified waste without inhibiting the power industry.[3]

ERDA has acknowledged the importance of broad public participation in the selection of technological options, so that the options ultimately chosen will gain broad public acceptance.[4]

NRC has set forth the broad goals of its regulatory efforts in these terms:

1. Isolation of radioactive wastes from man and his environment for sufficient periods (in some cases hundreds or thousands of years) to assure public health and safety and preservation of environmental values.
2. Reduction to as low as reasonably achievable:
   a. Risk to the public health both from chronic exposure associated with waste management operations and possible accidental releases of radioactive materials from waste storage, processing, handling or disposal.
   b. Long-term social commitments (land-use withdrawal, resource commitment, surveillance requirements, committed site proliferation, etc.)[5]

NRC has also expressed interest in attaining certain social, economic, and environmental goals. A recent NRC task force report on radioactive waste goals and objectives also reflects concern with the institutional arrangements required for waste management, economic impacts, the foreclosure of future options, time frames for action, distribution of hazards and benefits, the uncertainties permeating decision making, and public involvement in waste management decisions.[6]

In the context of the Carter administration's new nuclear power policy, these waste policy goals seem to be as appropriate for the management of unreprocessed spent fuel as for the management of reprocessing wastes.

Organizational Overview

Institutional arrangements are required to carry out waste management policy. The overall organization must include capacities for three generic functions: waste management itself, regulation of waste management to assure safety, and research and development to improve both management and regulation.

We use the term "waste management" to mean the actual performance of physical and administrative tasks that constitute the proper short- and long-term handling of radioactive materials. These activities include temporary storage, treatment, packaging, transportation, and retrievable storage or permanent disposition. Retrievable storage and permanent disposition may be termed "long-term" activities; all others are "short-term," regardless of when they occur. Current management responsibilities can be described by considering in turn each of the four categories of post-fission waste, based on type and origin, that lie within the scope of this report: (1) commercial HL waste; (2) military HL waste; (3) commercial TRU waste; and (4) military TRU waste. As we shall see in some detail subsequently, the responsibility for management is variously divided between ERDA and private industry depending on the type and origin of post-fission wastes.

The regulatory function includes primarily three types of activities: setting general criteria and specific standards, licensing or approving various managerial activities (e.g., siting, construction, and operation of facilities), and monitoring and enforcement related to ongoing operations. Again depending on the type and origin of post-fission wastes, primary responsibility for regulation is divided among NRC, ERDA (through self-regulation or regulation of its contractors), and certain state govenments. Supplementing the primary agencies are the regulatory responsibilities of several "nonnuclear" agencies, which derive their jurisdiction from legislation on transportation or the environment.

Contrasted with the complexity of institutions involved in management and regulation, a single agency—ERDA—is responsible for formulating and conducting or sponsoring the great bulk of radioactive waste research and technology development.

Management

Under the Atomic Energy Act of 1954, as amended,[7] and the Energy Reorganization Act of 1974,[8] waste management responsibilities are divided among ERDA and licensees of NRC or Agreement States.[9] ERDA personnel, however, do not actually operate ERDA's waste management facilities: this work is done almost entirely by ERDA contractors, subject to ERDA management policies. Thus, when we refer to ERDA or federal "management," we mean such a two-tiered managerial arrangement.

Commercial HL waste is to be managed by licensees until it is delivered to ERDA facilities for final disposition. Under current regulations licensees must solidify liquid HL waste no later than five years after its generation and must transfer it to an ERDA repository no later than five years after its solidification.[10] Thus private industry is responsible for the short-term (temporary storage, treatment, packaging, transportation), and the federal government is responsible for the long-term (retrievable storage or permanent disposition).

Responsibility for management of military HL waste resulting from weapons production and naval nuclear propulsion programs is comprehensively vested in the federal government. The Department of Defense (DOD) is a relatively minor manager of military waste, since it must handle spent fuel elements from naval reactors for only limited periods of time, until they are transferred to ERDA. DOD does not manage HL waste from weapons production, since ERDA itself produces plutonium for weapons and manufactures the warheads. As discussed in Chapter 3, it is during plutonium recovery that weapons HL waste is generated.[11]

The present law and NRC regulations allow private industry licensees to have both short- and long-term management responsibilities for commercial TRU waste. However, regulations proposed by the former Atomic Energy Commission (AEC) and inherited as proposals by NRC would alter this situation for the long-term. The proposed regulations could prohibit burial of any transuranic waste and require licensees to transfer all transuranic waste to federal (i.e., ERDA) facilities within five years of generation.[12] Resolution of certain technical and economic problems in the proposal, plus the preparation of an Environmental Impact Statement on it, are necessary before it becomes effective. All but one of the six commercial low-level waste burial sites now in operation have suspended TRU waste burial.[13]

Most military TRU waste is managed within the federal government by ERDA in the long-term and by ERDA or DOD in the short-term. Again, DOD manages relatively small amounts of TRU waste from weapons component assembly and naval propulsion programs, which it generally transfers to ERDA as soon as possible. In addition, several commercial operators inherited some buried military TRU waste when they took over sites which had previously been run by the AEC. At least one commercially operated burial site (at Hanford) continues to receive military TRU waste from nuclear navy operations.[14]

Regulation

The body of radioactive waste regulation is still in the early stages of development. This should be borne in mind throughout the following discussion.

NRC efforts now under way in the area of waste regulation include:

1. development of a regulatory program for radioactive wastes;
2. establishment of a confirmatory research program to support regulatory and licensing activities in waste management;
3. classification of the types of nuclear wastes;
4. development of specifications for the solids which will be acceptable forms for HL waste;
5. review of requirements for the disposal of low-level waste at commercial burial grounds; and
6. coordination of activities with state and local governments and other federal agencies.[15]

Criteria and Standards

Authority to promulgate radiation safety criteria and standards applicable to radioactive material is vested in the federal government.[16] NRC has authority, derived from the Atomic Energy Act of 1954 and the Energy Reorganization Act of 1974, to establish criteria and standards for protection against radiation applicable to licensed activities.[17]

ERDA develops criteria and standards for its own operations. EPA also has some authority in this area of regulation. When it was created in

1970, EPA received, among other powers, the authority previously vested in the Federal Radiation Council (which was abolished) to

> advise the President with respect to radiation matters, directly or indirectly affecting health, including guidance for all Federal agencies in the formulation of radiation standards and in the establishment and execution of programs of cooperation with States.[18]

The 1970 Reorganization Plan also transferred from AEC to EPA the responsibility for

> establishing generally applicable environmental standards for the protection of the general environment from radioactive material. As used herein, standards mean limits on radiation exposures or levels, or concentrations or quantities of radioactive material, in the general environment *outside the boundaries of locations under the control of persons possessing* or using radioactive material.[19] [Emphasis added.]

EPA has relied to date upon this latter language, which excludes from EPA radiation standard setting those activities of both NRC licensees and ERDA that are located geographically within the possessor/user's facility. (Thus EPA has jurisdiction to set exposure standards for transport of wastes.) Some EPA statements indicate an attempt to expand the standard-setting authority of the agency to reach within the boundaries of NRC-licensed or ERDA facilities by reviewing NRC's and ERDA's own standards for protection against radiation under the authority of its "guidance" mandate.[20] This could create regulatory redundancy and conflict.

Licensing/approval activities are the central function of post-fission waste regulation, and they present the most complicated aspect of the present situation. Although NRC and ERDA have general licensing/approval duties within their respective areas of jurisdiction, several other agencies, within and outside the federal government, also have claims of varying strength for additional, occasionally overlapping, authority to license. Before discussing these additional agencies, we will outline current allocations of authority between NRC and ERDA for the various regulatory activities, aside from standard setting, and identify some of the uncertainties that exist in these allocations.

Siting HL and TRU Waste Facilities

Siting requirements adopted to date concern mainly land ownership. ERDA's practice is to purchase, rather than to lease, land for its use;

thus ERDA waste management facilities are and will be located on federally owned land. "[D]isposal of high-level radioactive fission product waste material will not be permitted on any land other than that owned and controlled by the Federal Government."[21] Current NRC regulations require applicants for low-level waste burial licenses to provide an environmental analysis of the proposed site,[22] and allow burial only on land which is owned either by the federal or a state government.[23] Reprocessing plants and temporary storage facilities for HL waste may be located on privately owned property.[24] Whereas TRU waste disposal can now take place on state or federal land, proposed regulations would limit permanent disposition of such waste to property owned by the U.S. government.[25] Further siting requirements are being developed by the NRC. The role of other federal and state agencies in siting is discussed further below.

Facilities Construction and Operation: HL Wastes

NRC has sole authority to license commercial HL waste management facilities and operations from the standpoint of radiological safety. Moreover, section 202(3) of the Energy Reorganization Act of 1974 also subjects any ERDA facility handling commercially generated HL waste to NRC licensing. It would therefore appear that ERDA research, development, and demonstration facilities using commercially generated HL waste are subject to NRC licensing.[26] ERDA is not, however, planning to use commercial HL waste in its pilot plants for demonstration of a permanent waste repository.[27] ERDA is now studying the problem of obtaining a suitable quantity and composition of waste for use in its projects designed to serve as demonstrations of long-term waste repositories.

As noted above, ERDA contractors are responsible for day-to-day management of military HL waste. These contractors are both supervised (in a managerial sense) and regulated (for safety assurance) by ERDA through the terms of its contracts. ERDA manual chapters on waste management, which give very general guidance,[28] standards for protection against radiation,[29] requirements for reporting of occurrences,[30] etc., are incorporated by reference into contract terms. ERDA approval of contractor operations is not open to public scrutiny, as in the case of NRC approval of private industry operations which it licenses. But this type of ERDA self-regulation is specifically authorized by the Atomic Energy Act of 1954.[31] The approach may be justifiable in some instances on national security grounds. But it should be recognized that for the short term, ERDA exercises "regulatory" as well as managerial control over its own military HL waste.

For the long term, section 202(4) of the Energy Reorganization Act gives NRC licensing authority over the following ERDA facilities:

> Retrievable surface storage facilities and other facilities authorized for the express purpose of subsequent long-term storage of high-level radioactive waste generated by the Administration, which are not used for, or are part of, research and development activities.[32]

Under this provision, NRC must license the long-term disposition of military HL waste. However, definitional problems with the language of section 202(4) have created uncertainty as to the scope of NRC regulatory authority.

The Natural Resources Defense Council (NRDC) challenged the government's construction of the Energy Reorganization Act in July, 1975, in a memorandum supporting NRC licensing of ten proposed new waste storage tanks to be built at Hanford and Savannah River.[33] ERDA took the position that these tanks would not be subject to NRC licensing under section 202 of the Act because "such waste storage facilities have not been considered in the past by the Congress to be the type of facility which would be utilized by the AEC or ERDA for long-term storage of high level radioactive waste."[34] The ERDA stance was based also on the argument "that it was the Congressional intent when the...Act was enacted...that section 202(4) apply only to waste facilities, the design characteristics of which were then unknown to the Congress and that those facilities when submitted to the Congress in the authorization process would be clearly characterized...as 'authorized for the express purpose of subsequent long-term storage.'"[35] The NRDC filed suit against ERDA in September, 1976, seeking a declaratory judgment on the licensing issue, as well as an injunction against the construction of the tanks.[36]

Subsequently, NRC concluded that "the projected ERDA facilities are not now subject to licensing under section 202(4)."[37] The commission premised this conclusion upon ERDA assurances "that the use of the facilities will extend less than twenty years"[38] and "that no part of the facilities will be used for long-term storage,"[39] as well as the expression of a like opinion by the Joint Committee on Atomic Energy of the Congress.[40] NRC made the continuing validity of its decision "contingent upon ERDA's keeping it informed of any material changes in plans for these facilities and a report on their use and plans for future use in years following completion of construction.''[41]

There is at least one congressional statement to the effect that section 202(4) of the Energy Reorganization Act should require NRC licensing of these new tanks, and that ''it was our intent that any new construction of waste storage facilities by ERDA, including those built according to an existing design, should be licensed by the NRC.''[42] As of February, 1977, the outcome of the NRDC legal action had not been judicially determined.

Under present law, military HL waste already in existence at the time the Energy Reorganization Act was enacted is not immediately subject to NRC licensing. Some of the waste in question has been stored at ERDA facilities since the 1940s. Critics have asserted that any period longer than twenty to thirty years constitutes long-term storage.[43] More significantly, ERDA has for some time been solidifying liquid HL waste in place within storage tanks in order to prevent further leakage. The result, as discussed in Chapter 2, is that removal of this solidified HL waste for treatment and transfer to NRC-licensed, long-term waste repositories may prove very expensive.[44] In the meantime, ERDA will retain regulatory authority over military HL waste in those tanks which existed prior to the passage of the Energy Reorganization Act.

Nevertheless, NRC's basic licensing authority over the permanent disposition of military HL waste, which includes the promulgation of safety criteria for acceptance of waste at a permanent repository, could give it substantial indirect control over at least some aspects of short-term ERDA management (e.g., form of waste and type of packaging for the ultimate disposal in NRC-licensed waste repositories). This would not, however, enable NRC to exercise short-term monitoring and enforcement with regard to ERDA wastes, nor to establish a schedule for ERDA waste treatment in preparation for transfer to a permanent waste repository.

As noted above, section 202 of the Energy Reorganization Act excludes from NRC licensing ERDA research and development activities involving ERDA-generated HL waste. On October 28, 1976, President Ford issued a major statement on nuclear policy in which he directed, among other things, that:

> the first demonstration depository for high-level wastes which will be owned by the Government be submitted for licensing by the independent NRC to assure its safety and acceptability to the public.[45]

This statement settled the question as to whether waste demonstration facilities using ERDA-generated HL waste are subject to NRC licensing. ERDA officials have suggested that TRU waste might be used, rather than HL waste, in the initial waste repository demonstration which is most likely to occur in a salt formation near Carlsbad, New Mexico. Potential health and safety risks will be considered in such a demonstration and the results are intended to be relevant to the safety of long-term HL waste repositories, but President Ford's statement did not directly dispose of the issue of licensability if TRU waste is used in a demonstration long-term repository. The legal situation on this point is not clear.

Facilities Construction and Operation: TRU Waste

The 1959 amendments to the Atomic Energy Act of 1954 provided a means whereby states can regulate commercial low-level waste operations within their boundaries under the Agreement States program.[46] Pursuant to these provisions, NRC relinquishes this part of its regulatory power to states which desire this authority and conform to the commission's standards. But NRC retains the authority to suspend or revoke an agreement with a state if the public health or safety is threatened.[47] The suspension power has never been exercised thus far. Close working relationships between the states and the NRC are typical. There are now 25 agreement states. Pursuant to the Agreement States program, certain states wholly or partially regulate five of the six currently operating commercial low-level burial facilities: Barnwell, South Carolina; Beatty, Nevada; Hanford, Washington; Moorehead, Kentucky; and West Valley, New York. (The Sheffield, Illinois, site is under exclusive federal jurisdiction.)[48] Of course, even in agreement states, section 274 of the Atomic Energy Act reserves to NRC exclusive jurisdiction to license significant quantities of special nuclear materials.

Proposed NRC regulations would prohibit shallow land burial of TRU waste and shift long-term management of such waste from private operators to ERDA. If finally adopted, these regulations would take this aspect of regulatory authority out of the agreement states and place it once more in the federal government. It would be logical for NRC to have this role, in light of its parallel long-term regulatory responsibilities for HL waste. However, existing law expressly requires NRC licensing of long-term storage of HL waste only.[49]

NRC is considering simplifying the defined categories of radioactive waste into high- and low-level only,[50] assimilating what is now called transuranic contaminated low-level waste into the high-level category. Alternatively, NRC might attempt to exercise indirect control over long-term TRU waste repositories by including safety criteria for them in its requirements for transportation of TRU waste to the repositories by NRC licensees. A more straightforward solution to this ERDA–NRC jurisdictional problem would seem to require legislation.

NRC is rethinking the radioactive waste aspects of its Agreement States program. Discussion between NRC staff members and state government radiation control officials has begun.[51] Such reconsideration is taking place in an environment of waning state interest and financial capacity in the development of an adequate radiological safety program and increasing controversy about the safety of low-level waste burial sites. Apparently many states would willingly rid themselves of the expense and responsibility of this phase of radioactive waste regulation.

ERDA has sole regulatory authority over military TRU waste. DOD regulates the short-term management of TRU waste generated by weapons component assembly or naval nuclear propulsion programs at DOD facilities themselves, but this waste is transferred eventually to ERDA.

No definition of "high-level radioactive waste" is supplied by the Energy Reorganization Act itself. However, at the time the Act was passed, existing regulations excluded TRU waste from the high-level category.[52] Under current practices and proposed regulations transuranic waste is viewed with considerable concern since, as discussed in Chapter 2, the quantities of plutonium contained in low-level and high-level wastes generated by commercial plutonium recycle will be roughly equal.[53] Nevertheless, as we have seen above, the existing U.S. regulatory structure does not provide firm assurance of NRC regulation of either commercial or military TRU waste.

Transportation

Charged by a series of legislative enactments culminating in 1975, the Department of Transportation (DOT) shares responsibility with the nuclear agencies for assuring safe transport of hazardous radioactive materials in commerce.[54] Recognizing their common jurisdiction, DOT and the former AEC in 1973 entered into a Memorandum of Understanding which allocates regulatory duties in this area.[55] The Memorandum recognizes DOT's general authority to develop safety standards for packaging by shippers and handling by carriers, but it authorizes AEC to develop criteria for packages of Type B, of large quantity, and of fissile materials. Each agency further pledges to enforce the other's standards and to exchange information prior to issuing new regulations in this area. As noted, the Memorandum was signed before the AEC functions were split between NRC and ERDA. It is now applicable to both federal nuclear agencies, since the former AEC specifically imposed the same standards of care for transport upon both its licensees and its license-exempt contractors. ERDA has acknowledged its obligations by incorporating the relevant AEC manual chapter into its own body of manual chapters, which is incorporated by reference into all ERDA shipper contracts.[56] NRC likewise has adopted without significant variation that part of previous AEC regulations pertaining to the packaging of radioactive material for transportation by its licensees.[57] Both nuclear agencies also require shippers to follow a registration and approval procedure that parallels DOT regulations for the shipment of nonradioactive hazardous materials.[58] As indicated above, all waste shipments outside ERDA or NRC facilities must also meet EPA's generally applicable standards for protection of the environment against radiation.[59]

There is one exception to the federally dominated scheme of waste transport regulation. In the case of transportation of waste containing less than critical mass quantities of special nuclear materials within agreement states, neither DOT nor NRC has jurisdiction. State laws and regulations apply instead. However, all agreement states require packaging to comply with DOT standards.[60]

Current federal transportation and packaging regulations, in effect since 1968, are virtually identical to the regulations adopted by the International Atomic Energy Agency (IAEA) in 1961. The IAEA, however, revised its regulations in 1973, and ERDA and NRC are considering revisions in response to these changes by the international organization.

A number of recent radioactive waste transportation incidents have sparked controversy. Also, most of the state nuclear power initiatives that have been proposed include provisions on transportation. In response, NRC has committed itself to a program of emergency response planning, environmental impact assessment, and risk analysis.[61]

Monitoring and Enforcement

The Atomic Energy Act of 1954 grants both NRC and ERDA ample authority for monitoring and enforcement.[62] Monitoring consists mainly of screening required reports, auditing managers' inventories, and inspecting sites. Enforcement involves the punishment of violators and ordering remedial safety actions, particularly during emergencies. NRC has general regulations governing these activities,[63] and ERDA conditions its contracts on adherence to manual chapters with similarly general requirements.[64] If a contractor cannot be convinced to comply with manual chapter standards, ERDA may rescind the contract and, if necessary, confiscate the facility.

It is unclear, however, how ERDA manual chapters compare in legal weight with regulations such as those formulated by NRC and published in the Code of Federal Regulations. From ERDA's standpoint as a regulator, the question is unimportant, since it can exert contractual pressures on its managers. But the issue may be important to an intervenor seeking a court order against ERDA requiring it to enforce its own standards.

DOT has general monitoring and enforcement authority regarding radioactive waste transport.[65] Maritime shipments of wastes are subject to Coast Guard regulatory jurisdiction.[66] ERDA shipments are, however, exempt from Coast Guard regulatory jurisdiction if accompanied by ERDA personnel.[67] The Federal Aviation Administration's role in waste shipment regulation is negligible, since neither NRC nor ERDA contemplates the use of air shipment in waste management operations.

Federal–State Relations

The general role of individual states in waste management regulation is an issue of increasing importance. This issue underlies much of the previous discussion.

A number of states have already challenged the notion of exclusive federal jurisdiction over radioactive waste regulation. The numerous state nuclear power initiatives which have been proposed, all of which have been rejected so far by large majorities of the voters, have generally contained provisions expressly aimed at the radioactive waste problem. The California legislature enacted several laws intended to impose milder conditions on nuclear power development in the state than were contained in Proposition 15, which was defeated in June, 1976. One of these recently enacted laws prohibits any nuclear power plant land use in the state or certification by the State Energy Resources Conservation and Development Commission until

> the commission finds that there has been developed and that the United States through its authorized agency has approved and there exists a demonstrated technology or means for the disposal of high-level nuclear waste.[68]

Michigan's governor informed ERDA during the summer of 1976, in response to ERDA's preliminary investigation of possible disposal sites in the state, that

> before any further negotiations, discussions, or binding decisions or contracts are made, I, as Governor, must be assured in writing that a disposal site will not be selected without approval of the state.[69]

Also in 1976, Connecticut passed legislation prohibiting the transportation of commercial radioactive waste within or through its boundaries unless the Connecticut Commission of Transportation has first issued a permit authorizing its shipment. The law includes strong inspection and enforcement provisions, including civil fines of up to $10,000 for each violation.[70]

The federal government is constitutionally established as a government of limited powers. All authority not delegated in the U.S. Constitution is expressly reserved to the several states by the tenth Amendment. In general, the states retain broad authority to legislate and to regulate activities within their respective borders in order to assure public health and safety. Land use is also a subject that is traditionally considered within primary state control. Therefore, state regulation of radiological hazards in general, and radioactive waste in particular, would clearly be appropriate and necessary, but for two qualifications. The first qualification is that, under the Commerce Clause, state regulation must not constitute an undue burden on interstate commerce;[71] the second is that, under the Supremacy Clause, state regulation must not conflict with valid federal legislation or regulation.[72] Of the two, the preemption issue is more important in view of the existing federal legislation and regulatory scheme previously discussed. Based on the pervasiveness of congressionally mandated federal regulation of radioactive waste management and the congressional intent expressed in legislative histories of the relevant acts, it is reasonable to conclude that current federal law (including the Agreement States program) would preempt attempts to regulate radioactive waste operations for the purpose of controlling radiological hazards.[73] Such a conclusion is supported by *Northern States Power v. Minesota,*[74] in which the U.S. Supreme Court affirmed the U.S. Court of Appeals for the Eighth Circuit holding that the AEC-prescribed limits on radioactive effluents from nuclear power plants preempt state regulation, and that states cannot impose more stringent radiological standards.

However, state land use controls which in effect zone out radioactive waste operations from certain areas may still be valid, depending on the purpose of the exclusion and the specific character of the activity excluded. Probably a state or locality, acting pursuant to state law, could control the siting of reprocessing and fuel fabrication plants, including temporary radioactive waste storage facilities, as part of a broader scheme to control industrial development and the location of major industrial facilities.[75] State disapproval of a site for a nuclear facility must rest on grounds other than the potential radiological hazard involved. On the other hand, a state or locality probably could not prevent the siting of a permanent radioactive waste repository, which is legally required to be on federal land, presently owned or acquired.[76] Federal radiological regulation would appear to preempt state efforts to regulate transportation of radioactive waste to the extent the state's concern is based on safety issues. Finally, although it is clear that if there are nuclear power facilities in a state, that state cannot impose radiation protection standards different from those adopted by the federal government, a state law which prohibits outright the construction of any nuclear power plants in the state has not been tested in court as of February, 1977.

The operational reality of federal–state relations in any field is framed by the law, but it is influenced decisively by politics and economics. Whether or not state regulation would be eventually declared federally preempted and hence invalid in a court test, the governor, legislature, or people of a state can in many ways effectively resist an activity within the state's borders that is authorized or directed by the federal government. This is true of the location of military installations, such as the Navy's Project Sanguine (now Project Seafarer),[77] and of radioactive waste repositories, such as the AEC's defunct proposal for a repository in a salt formation near Lyons, Kansas.[78]

States interested in participating in the regulatory decision-making process will probably be able to exert their greatest influence at the site-approval stage of a federal licensing process. As long as state claims are not merely pretexts to bar these facilities outright, states must be allowed to assert reasonable influence in siting and other regulatory decisions that will affect the nonradiological safety and welfare of local communities. It is presently unclear how serious federal–state conflicts over regulatory jurisdiction with regard to radioactive waste will become. In any case, political pressures are likely to play at least as forceful a role as legal precedent in the ultimate resolution of whatever conflicts arise. There is already evidence of political bargaining in the development of solutions to the problem of repository siting,[79] and the federal nuclear agencies presently appear determined to work cooperatively with the states concerned.

International Regulation

Activities involving radioactive waste may occur within or beyond the limits of national jurisdiction. Moreover, radioactive waste operations within one nation may have effects beyond that nation's jurisdiction, either in another nation or in an area outside the jurisdiction of any nation, such as the high seas, the deep seabed, Antarctica, or in space. Radioactive waste management activities occurring or having effects beyond U.S. jurisdiction may, if conducted by U.S. nationals, be subject to U.S. regulation, or regulation by other affected nations, or various forms of international regulation, or all three concurrently. The international ramifications of radioactive waste are important to consider, especially in view of existing practices regarding ocean dumping of low-level waste by other nations and future options for seabed and ice-sheet disposal of HL waste.

Nationally, under the Marine Protection, Research and Sanctuaries Act of 1972,[80] EPA has authority to issue permits for the transportation of all materials (except dredged material, which is the jurisdiction of the

Secretary of the Army) for the purpose of dumping them in the territorial waters of the United States.[81] Dumping of HL waste is specifically prohibited,[82] but the dumping of TRU waste is not. The U.S. government has nonetheless enforced a moratorium on TRU waste dumping since 1970.[83] Neither EPA nor ERDA has yet interpreted "dumping" to include emplacement of radioactive waste under the ocean floor in the deep seabed. However, EPA has claimed that it has licensing jurisdiction over such action by U.S. nationals.[84] Both the meaning of dumping and EPA's assertion of jurisdiction are subject to dispute, and may require congressional or judicial clarification.

As a brief digression at this point, it is noteworthy that EPA has parallel authority to control fresh water pollution through a different permit-issuing program under the Federal Water Pollution Control Amendments of 1972.[85] Under the legislation, HL waste "discharge" is prohibited.[86] No one, of course, contemplates using U.S. fresh waters for post-fission waste disposal purposes, but the prevention of radionuclide migration from temporary storage facilities or permanent repositories to aquifers is one of the primary technological goals of radioactive waste management. EPA has no authority to license radioactive waste facilities on land. It has maintained that NRC and ERDA have sole jurisdiction over discharge of source, by-product, and special nuclear materials, a position recently affirmed by the Supreme Court.[87] It is not clear what effect, if any, this decision will have on EPA's claim for permit-issuing power over possible ERDA deep seabed emplacement, where, again, only source, by-product, and special nuclear materials are involved.

Internationally, the United States is party to a number of multilateral agreements relating to U.S. radioactive waste policy. Two have major significance. The 1958 United Nations Law of the Sea Convention on the High Seas,[88] applying to all parts of the sea not included in the territorial waters of a nation, provides the foundation for modern international attempts to protect the maritime environment. The High Seas Convention makes it the general duty of nations to exercise "reasonable regard to the interests of other states in their exercise of the freedom of the high seas."[89] Three more specific obligations apply directly to the problem of marine pollution: (1) nations are to promulgate their own regulations so as to prevent pollution "from the exploitation and exploration of the seabed and its subsoil";[90] (2) "taking into account" any international regulations, nations "shall take measures to prevent pollution of the seas from the dumping of radioactive waste";[91] and (3) nations have a duty to "cooperate with the competent international organizations [here, the IAEA] in taking measures for the prevention of pollution of the seas or air space above."[92] These High Seas Convention obligations are by no means stringent. Obviously, the United States and other nations can choose to interpret the "reasonable regard" standard strictly or loosely and then behave accordingly. The requirements to take into account international regulations and to cooperate with IAEA, however, imply at least consultation with IAEA prior to implementing deep seabed emplacement of HL waste. (Ocean dumping of TRU waste is apparently not considered an acceptable U.S. option, although other nations are continuing to dump low-level waste, and it is not clear whether some of this waste is TRU contaminated.)

In 1975 the London Convention on Prevention of Marine Pollution by Dumping of Wastes and Other Matter[93] entered into force. It prohibits, except in the case of emergencies at sea and other emergencies for which no other solution is feasible, "dumping" of HL waste at sea.[94] An appropriate national authority is required to monitor ocean dumping of low-level radioactive waste; reports on each dump must be made to it.[95] Regional cooperation is encouraged, but enforcement is specifically a national matter.[96] The "emergency" exceptions have been criticized as possibly allowing parties a means of exempting themselves from the dumping prohibition. Although the parties probably did not intend "dumping" to include deep seabed emplacement, the issue is not closed, because of subsequent treaty interpretations.

The United States is also a party to the 1959 Antarctica Treaty,[97] which specifically prohibits the "disposal" of radioactive waste in Antarctica.[98] Each of the parties is entitled to designate observers, who have complete freedom of access at all times subject only to the jurisdiction of their nation, to monitor compliance with this ban, among other obligations in the Treaty.[99] The parties further agree "to exert appropriate efforts, consistent with the Charter of the United Nations," to ensure that no one engages in radioactive waste disposal in the Antarctic.[100] Hence, use of the Antarctic ice sheet as a permanent repository for HL waste is clearly prohibited for the parties to the treaty. However, in 1991 the treaty becomes subject to review and possible modification on the request of any party,[101] and the ice-sheet disposal option may be considered in negotiating a modification.

The outcome of the current Law of the Sea Conference may also affect the prospects for seabed disposal. The Informal Single Negotiating Text, which has formed the basis of the negotiations, contains provisions which would clarify and reinforce existing international legal norms affecting marine pollution generally. The Text would establish an International Seabed Authority to regulate development of seabed resources. Although primarily concerned with resource development, the competence of the authority would seem necessarily to extend to seabed disposal of radioactive waste, at least to the extent of assuring that any proposed waste disposal activity does not unduly interfere with existing or possible future seabed resources exploration, development, or production activity.

Finally, other governments may specially object to any disposal by the United States of radioactive waste of military origin in an area beyond its national jurisdiction. The motive for such an objection may be general concern with the impact on the environment, or it may be political concern with the military origin of the waste involved.[102]

In view of the international constraints regarding radioactive waste operations noted above, it is noteworthy that, as part of his October 28, 1976, statement on nuclear policy, President Ford directed the Secretary of State:

> to discuss with other nations and the IAEA the possibility of establishing centrally located, multinationally controlled nuclear waste repositories so that the number of sites that are needed can be limited.[103]

Such a proposal could well have advantages for a number of industrial countries with large nuclear power programs which lack suitable sites for waste repositories on their national territories (for instance, Japan and some European countries), and also for developing countries which do not need to develop their own facilities for the back end of the nuclear fuel cycle because of the relatively small size of their nuclear power programs. Moreover, such a proposal for multinationally controlled radioactive waste repositories would supplement President Ford's proposal in the same policy statement for a "new international regime to provide for storage of civil plutonium and spent reactor fuel," and it would complement previous U.S. proposals for the establishment of multinational nuclear fuel centers as a means of reducing the risks of nuclear weapons proliferation.

Research and Development

The Energy Reorganization Act left ERDA with principal research and development authority for military and commercial radioactive waste technology.[104] NRC and EPA conduct essentially confirmatory research and studies aimed at improving the quality of their regulations. NRC recognizes ERDA's broader research capabilities and must necessarily rely on much of the data furnished by ERDA without confirmation. Such a separation is important so that NRC, as an independent regulatory agency, does not find itself in a position where it is being asked to license technology developed by itself.[105]

ERDA's Fiscal Year 1977 budget reflected the dramatically growing governmental concern over finding technological answers for radioactive waste management problems. The combined military and commercial programs received a three-fold increase in funds from approximately $30 million to about $90 million.[106] Commercial waste programs within this budget were expanded by a factor of 5, from about $12 million to over $60 million.[107] ERDA summarized the priorities in its budget as follows:

> The ERDA FY 1977 commercial waste budget is keyed to providing input decisions on reprocessing, waste forms and storage modes for high-level radioactive waste. The selection of specific sites and development of repositories for terminal storage is considered the major item in a greatly expanded ERDA Fiscal Year 1977 waste budget. For earlier necessary or desirable waste management activities, technology is either in use or considered well developed, i.e., the basic technical principles are clearly understood and data needed for design are available. Efforts to reduce these principles to practice, especially in high-level liquid waste solidification, are also the subject of expansion in the FY 1977 budget. All of the foregoing effort has been keyed to the expected needs of the nuclear fuel cycle industry.[108]

ERDA, then, plays the role of principal researcher, in addition to its roles as major manager and self-regulator of most of its own management activities. Since "research and development" as used in the Atomic Energy Act include basic authority to engage in demonstration projects,[109] ERDA also has the authority to demonstrate the feasibility of ultimate deep geologic disposal (the currently favored technology), as well as other phases of radioactive waste operations. ERDA decisions made in the research and development area and also with regard to demonstrations involving TRU waste seem insulated from NRC regulatory control. Yet, such decisions profoundly affect the direction of the national waste management effort and policy making in this area.

Implementation

In 1976, the Federal Energy Resources Council, in cooperation with several agencies, including ERDA and NRC, issued a status report on radioactive waste management, in which it listed four requirements for proper implementation:

The thorough reviews mandated by the National Environmental Policy Act,
The promulgation and satisfactory compliance with generally applicable environmental standards and criteria issued by the Environmental Protection Agency,
Compliance with licensing criteria and requirements of the Nuclear Regulatory Commission, and
Opportunities for full public participation.[110]

The National Environmental Policy Act[111] (NEPA) is an omnipresent federal policy-making mechanism. It requires every federal agency to prepare an Environmental Impact Statement (EIS) in connection with "every recommendation or report on proposals for legislation and other major Federal actions significantly affecting the quality of the human environment."[112] Each EIS must contain a detailed statement by the responsible officials on:

the environmental impact of the proposed action,
any adverse environmental effects which cannot be avoided should the proposal be implemented,
alternatives to the proposed action,
the relationship between local short-term uses of man's environment and the maintenance and enhancement of long-term productivity, and
any irreversible and irretrievable commitments of resources which would be involved in the proposed action should it be implemented.[113]

It must be prepared in consultation with and must include the comments of "any Federal agency which has jurisdiction by law or special expertise with respect to any environmental impact involved."[116]

NEPA does not require avoidance or rectification of environmental harm from a proposed project. But any EIS not prepared in accordance with the procedural requirements set forth above may be deemed "inadequate" by a court. An agency may then be required to draft a new EIS. EIS's, when published, become part of the public record. EPA, sometimes in conjunction with the Council on Environmental Quality, reviews and comments on all EIS's.[115] Public sentiment and/or EPA ratings can be important factors in congressional decisions to fund agency proposals or in court hearings on whether or not to enjoin agency actions.

Operations in the back end of the nuclear fuel cycle which are federally licensed or contracted for constitute "federal actions," as does the promulgation of regulations designed to control these operations. The question of which contemplated actions are "major" and likely to "significantly" affect the environment is more complicated. But it is clear that, prior to implementation, ERDA and NRC must prepare "generic" EIS's for their respective management and regulatory programs in regard to radioactive waste. Thereafter, ERDA must prepare an EIS in connection with each major step it takes to implement its waste management program regarding either military or commercial waste, and NRC must prepare an EIS specifically covering every license it issues for radioactive waste operations, whether the proposed operation is to be conducted by ERDA or private industry. On the horizon is the EIS which will be necessary for the planned pilot federal HL waste repository.

As discussed in Chapter 2, HL waste emerges in the course of nuclear fuel reprocessing. Before commercial processing can begin, NRC must complete its Generic Environmental Impact Statement on the Use of Recycle Plutonium in Mixed-Oxide Fuel in LWR's (GESMO), complete an extensive public hearing on the subject, issue final regulations governing the licensing of commercial reprocessing plants, and complete an EIS, hearings, and issue a license for the operation of a particular plant. The health, safety, and environment section of GESMO was completed in August, 1976. The safeguards supplement was to have been issued in draft form in early 1977.

In May, 1976, federal decision making concerning reprocessing and recycling suffered a legal setback in *NRDC v. NRC*.[116] The NRC had announced circumstances under which it might in the future have decided whether or not to permit interim licensing of a mixed oxide fuel fabrication plant to occur. The U.S. Court of Appeals for the Second Circuit concluded that any interim license of a fuel fabrication or reprocessing facility would inevitably be a substantial and irretrievable commitment to wide-scale use, and so could not occur without completion of the full GESMO process. It found that the NRC interim licensing procedure failed to satisfy NEPA specifically because of the inadequate treatment of alternatives to plutonium recycle and of hazards of theft, diversion, and sabotage. The U.S. Supreme Court subsequently granted *certiorari* in order to review this decision, but the Carter administration's new nuclear power policy may well have rendered the case moot.

Following the administration's decision on April 7, 1977 to defer indefinitely commercial reprocessing in the U.S., the GESMO hearing board stopped further proceedings. Moreover, at this writing, the safeguards supplement remains unissued. Thus, an unfortunate consequence of the deferral of reprocessing may be a substantial delay in the development of an effective system of safeguards against theft of materials that might be used to make nuclear explosives.

Legal developments during the summer of 1976 made radioactive waste an important unresolved issue in the commercial power reactor licensing process. The situation is worthwhile describing in some detail because of the insight that will be gained into the problem of regulatory uncertainty which currently afflicts the nuclear power industry.

In July, the U.S. Court of Appeals for the District of Columbia Circuit ordered NRC to reconsider the Vermont Yankee reactor operating license and the procedure used to issue the license.[117] Prior to the decision in *Vermont Yankee,* the issuance of power reactor operating licenses had been based, in part, upon an AEC rule which concluded that waste hazards in reactor operations were "insignificant." The court found this determination "arbitrary and capricious," because the factual premises upon which it was based had not been adequately supported or "ventilated."

Shortly thereafter, the NRDC filed a petition with NRC requesting the adoption of rules for implementation of the *Vermont Yankee* decision.[118] On August 16, the commission announced the reopening of the rule-making proceeding on the Environmental Effect of the Uranium

Fuel Cycle.[119] This rule making was reopened for the purposes of supplementing the record on the reprocessing and waste management issues, and determining whether or not, on the basis of the supplemented record, the disputed rule should be amended, and if so, in what respect. NRC suspended issuance of any new full-power operating licenses, construction permits, or limited work authorizations until the promulgation of an interim rule on the matter. As a basis for the reopened proceedings, NRC completed, in October, 1976, a new environmental survey of the probable contributions to the aggregate environmental effects of a nuclear power reactor that are attributable to the reprocessing and waste management stages of the fuel cycle.[120] At the same time, a proposed interim rule was announced,[121] and the suspension of licensing was lifted shortly thereafter. On the basis of the environmental survey, and drawing on comments received on the survey and the proposed interim rule, a "final interim rule" will be promulgated in 1977.

In parallel with these developments, NRC requested and obtained a judicial stay of the court's mandate, which allowed NRC to resume the granting of licenses for power reactors, but such licenses are contingent on the final outcome of the legal proceedings. The promulgation of a final rule will be preceded by procedures more demanding than the notice-and-comment procedures used to formulate the "final interim rule," but it remains to be seen exactly what sort of consideration of the reprocessing and waste management issues will be deemed adequate for the purposes of compliance with NEPA. In the meantime, electric utilities are able to apply for power reactor licenses—but at their own risk. So far, four nuclear power plants have begun operation with conditional licenses. To cap it all, in early 1977 the U.S. Supreme Court agreed to review the original July 1976 decision of the U.S. Court of Appeals in the *Vermont Yankee* case which invalidated NRC's licensing procedures.

In a separate action, NRDC filed a petition with NRC requesting the commission to make a determination under the Atomic Energy Act of 1954 that waste generated in reactors will be disposed of safely in the future before it issues operating licenses to new reactors.[122] As of December, 1976, the outcome of this action had not been resolved.

Procedures analogous to those described above with regard to commercial HL waste (EIS's, hearings, rule making, licensing) must also occur for the government's military HL waste program. The NRDC complaint on the new waste tanks noted above alleged in part that

''ERDA has determined not to prepare environmental impact statements under NEPA with respect to [the Hanford and Savannah River projects] and has announced its decision in written negative declarations.''[123] ERDA, on the other hand, has said that it will include discussion of the environmental impact of these tanks in its documents describing waste operations at these facilities. The administration has announced a timetable to complete, by the fall of 1977, a set of three defense waste documents—for Savannah River, Idaho Falls, and Richland—which will describe the technical alternatives available for future management of the HL waste at these sites.[126] The defense waste documents will provide the technical foundation for the environmental impact statements that will ultimately be prepared for long-term military HL waste management.

The NRDC recently petitioned NRC to adopt interim regulations governing disposal of low-level radioactive waste (including TRU waste) and to prepare a programmatic EIS on this aspect of waste management.[125] As of December, 1976, NRC had not issued a disposition of this petition. NRC's waste classification effort could conceivably redefine HL waste to include some or all TRU waste. In any case, NRC promulgation of the 1974 AEC-proposed TRU waste regulation must await action on the above petition.

Prior to President Ford's statement on nuclear policy of October 28, 1976, ERDA's time schedule for implementing commercial waste management technologies was as follows:[126]

1979—technology for handling commercial post-fission waste, other than high-level, available for industrial adoption;
1983—high-level waste solidification technology available for startup of commercial plants;
1984—technology available for packaging, transporting, and handling spent fuel, in the event of a decision not to reprocess spent fuel;
1985—at least one repository ready to receive waste.

President Ford's nuclear policy statement directed ERDA:

> to demonstrate all components of waste management technology by 1978 and to demonstrate a complete repository for such wastes by 1985.

As noted, however, President Carter has announced a major change in U.S. domestic nuclear energy policy. The President stated, on April 7, 1977, that ''[the United States] will defer indefinitely the commercial reprocessing and recycling of the plutonium produced in the U.S. nuclear power programs.''

The ultimate effect of this deferral on the waste management implementation schedule remains to be seen. Nevertheless, the radioactive waste of military origin, which, as noted in chapters 1 and 2, will in any event present the most difficult and costly technological problems to be resolved, is obviously unaffected by the policy change. Furthermore, the deferral decision increases the need for consideration of the management and storage of spent power reactor fuel. Finally, in view of both the possibility of a future reversal in U.S. domestic reprocessing policy and the interests of other nations in pursuing their reprocessing objectives, it would seem only prudent for the U.S. to continue the waste management program to a point at which the capability to treat, transport, and dispose of reprocessing plant wastes is satisfactorily demonstrated.

Conclusions

The U.S. inventory of radioactive waste has two sources—commercial and military nuclear activities. Military wastes presently constitute by far the greatest proportion of the most toxic waste. The future management challenges are comparable in their overall importance, although the difficulties, costs, and schedules differ.

1. In their statements of basic policy goals, federal officials have thus far adopted a primarily technological perspective. ''Safe management'' has essentially meant ''isolation and containment.'' More recently, policy makers have mentioned, however tangentially, other factors such as risk/cost analysis, institutional stability, public confidence, consolidation of management responsibilities, and interagency and international coordination.

2. The present organization for achieving radioactive waste policy goals has basically three functions: management, regulation, and research and development. Overall management responsibility is divided in a complicated way between the private industry (licensees of NRC and/or agreement states) and the federal government (ERDA and its contractors). Commercial HL waste is to be managed in the short term by licensees and in the long term by ERDA. Military HL waste is solely

a federal responsibility: ERDA manages plutonium production facilities that create HL waste, and DOD is responsible for managing spent fuel elements from naval propulsion programs until they are turned over to ERDA. Commercial TRU waste is presently managed by licensees in the short and the long term. NRC has proposed regulations, however, that will shift long-term management responsibility to ERDA. Present military TRU waste is managed solely by the federal government, primarily by ERDA.

3. The landscape of regulatory authority appears rather cluttered. It is composed mainly of NRC and ERDA at the level of federal government, but other federal agencies, state governments, and international bodies also have regulatory claims, though some are uncertain. Primary authority for setting criteria or standards for radioactive waste operations rests with NRC and ERDA in their respective areas of regulatory responsibility. EPA intrudes on this scene, however, with its authority to set generally applicable environmental standards for radiation exposure and to give guidance to all federal agencies in the formulation of radiation standards.

The heart of regulatory activity in waste operations is the licensing/approval function. Responsibility for licensing commercial HL waste operations rests solely with NRC. This authority appears to cover, indirectly, any ERDA research and development storage facility that receives commercial HL waste.

Military HL waste management is approved in the short term by ERDA, which conducts its monitoring and enforcement activities through contractual, rather than regulatory, mechanisms. For the long term, NRC has licensing authority encompassing military HL waste storage facilities not used for research and development, excluding ERDA facilities existing at the time the Energy Reorganization Act of 1974 became effective. Thus ERDA retains regulatory authority over its then existing facilities despite the continued absence of plans to transfer the military HL waste solidified in place at Hanford and Savannah River to ERDA-planned, NRC-regulated permanent repositories.

Also, ERDA approves its own site selection, construction, and operation of research and development storage facilities. However, NRC will license ERDA demonstration repositories for HL waste, at least if this type of waste is involved in the demonstration.

Regulatory authority over commercial TRU waste is exercised by NRC or relinquished, wholly or in part, by NRC to agreement states.

The present scheme would be altered, however, by proposed NRC regulations requiring all short-term managers to transfer their TRU waste to ERDA for long-term management; ERDA would then acquire long-term regulatory authority. If these regulations are adopted, NRC could, if it desired, implement a short-term regulatory strategy which might effectively give it control over long-term regulation. Such a strategy, however, would not infringe upon ERDA's current regulatory jurisdiction over military TRU waste, where ERDA has undisputed short- and long-term authority.

Overlapping the regulatory authorities of NRC and ERDA is DOT's responsibility to ensure the safe transport in commerce of hazardous radioactive materials. Cooperative arrangements between DOT and AEC (applicable now to both ERDA and NRC) divide responsibilities for establishing criteria for packaging and handling waste and allot sole responsibility for licensing of waste transport to the nuclear agencies, who agree to enforce their respective standards and applicable DOT ones as well.

4. The states' authority to regulate waste management operations is basically preempted by the federal government. Though challenges are continuing, the existing law is that states may only regulate non-radiological aspects in a way that does not unduly interfere with federal promotion and regulation of nuclear energy. Political pressures from concerned states can, however, be quite effective in altering the federal government's behavior. There is also ample room for state (and citizen) participation in federal decision processes.

5. Ocean-use options involve additional national and international environmental institutions. Nationally, EPA has permit-issuing authority for ocean dumping of TRU waste, though the United States does not currently employ such an option. Dumping of HL waste is prohibited, though probably this would not include deep seabed emplacement. EPA has also claimed permit-issuing authority over any deep seabed waste emplacement. Internationally, formal agreements to which the United States is a party place no unreasonable constraints upon U.S. radioactive waste decisions, but they do require a balancing of other nations' interests. High-level waste "dumping" is prohibited (with "emergency" exceptions), though here again it is doubtful that the term includes deep seabed emplacement. No effective international monitoring or enforcement institution currently exists. The outcome of the Law of the Sea Conference, which is presently unclear, must be taken into

account if the U.S. government decides to pursue any of the "ocean options" for radioactive waste disposal. Finally, Antarctica is legally off-limits for any type of radioactive waste disposal.

6. ERDA dominates waste management research and development. NRC and EPA contribute mainly confirmatory research in areas of particular regulatory interest. ERDA thus occupies the dual position of chief researcher and major manager of high- and low-level waste (as well as that of self-regulator of much of its own military waste management activities). A potential problem with such an institutional arrangement is that the managers within ERDA find themselves forced into management strategies dictated by their research counterparts, instead of being able to direct research efforts according to the necessities of waste management operations. On the other hand, there is a need for close coupling of research and development to operations, especially during the early stages of technological innovation.

7. In view of recent basic changes in U.S. nuclear policy which will defer indefinitely the commercial reprocessing of power reactor fuel, there is an urgent need to address the problems of large-scale management and storage of spent fuel. Moreover, a schedule for dealing with existing military waste remains to be developed. It will also be desirable to proceed with federal waste management programs for commercial waste to the point of demonstration. In all cases, policy decisions regarding radioactive waste are not self-executing. Implementation depends on the existence of a dedicated organization, the availability of adequate technology and funds, and—last but not least—public acceptability of proposed actions.

Existing institutional arrangements for management and regulation of radioactive waste are inadequate for the future.

Notes to Chapter 3

1. *Hearings on Radioactive Waste Management Before the Subcommittee on Environment and Safety of the Joint Committee on Atomic Energy,* 94th Cong., 2d Sess., Vol. 3, 74 (1976) (stenographic transcripts); hereinafter cited as "May hearings."
2. *Hearings on Storage and Disposal of Radioactive Waste Before the Joint Committee on Atomic Energy,* 94th Cong., 1st Sess., 18 (1975); hereinafter cited as "November hearings."

3. Address by Dr. Carl W. Kuhlmann, Assistant Director for Waste Management, Division of Nuclear Fuel Cycle and Production, ERDA, Atomic Industrial Forum Uranium/Nuclear Fuel Conference, March 21–24, 1976.

4. May hearings, Vol. 1, 34.

5. Memorandum from Kenneth Chapman, Director, Office of Nuclear Materials Safety and Safeguards for the NRC Commissioners, "Status Report—Waste Management Program," April 22, 1976, 6.

6. *Ibid.* at 3, 4.

7. Atomic Energy Act of 1954, 42 U.S.C. §§ 2011–2296 (1970).

8. Energy Reorganization Act of 1974, 42 U.S.C. §§ 5801–5891 (Supp. IV, 1974).

9. See text at note 45.

10. 10 C.F.R. § 50 App. F2. (1971) (policy relating to the siting of fuel reprocessing plants and related waste management facilities).

11. ERDA manufactures the nuclear explosive components of nuclear weapons, which DOD then assembles.

12. 39 *Fed. Reg.* 32921 (1974).

13. *Hearings on Low-Level Radioactive Waste Disposal Before a Subcommittee of the House Committee on Government Operations,* 94th Cong., 2d Sess., 209 (1976); hereinafter cited as "March hearings."

14. Interview with Robert Will, Director, Division of Radiation Control, State of Washington, in Olympia, Wash., July 20, 1976.

15. U.S. NRC Release No. 76-98, April 23, 1976, 2,3.

16. 42 U.S.C. §§ 2073, 2093, 2095, 2111, 2133, 2134, 2201 (1970).

17. *Ibid.,* 42 U.S.C. §§ 5814(c), 5841(f) (Supp. IV, 1974).

18. 42 U.S.C. § 2021(h) (1970); transferred by Reorganization Plan No. 3 of 1970 § 2(7), 42 U.S.C. § 4321 (1970).

19. Reorganization Plan No. 3 of 1970 § 2(6), 42 U.S.C. § 4321 (1970).

20. Address by William Rowe, EPA, at International Symposium on the Management of Wastes from the LWR Fuel Cycle, Denver, Colo., July 12, 1976.

21. 10 C.F.R. § 50 App. F3 (1971).

22. 10 C.F.R. § 20.302(a) (1971).

23. 10 C.F.R. § 20.302(b) (1971).

24. 10 C.F.R. § 50 App. F1 (1971).

25. See footnote 12.

26. 42 U.S.C. 5842(3) (Supp. IV, 1974).

27. Interview with Alex Perge, Office of Waste Management, Division of Nuclear Fuel Cycle, ERDA, in Germantown, Md., June 28, 1976.

28. ERDA Manual Chapter 0511.

29. ERDA Manual Chapter 0524.

30. ERDA Manual Appendix 0502.

31. 42 U.S.C. § 2140 (1970).

32. 42 U.S.C. § 5842(4) (Supp. IV, 1974).

33. Brief for Natural Resources Defense Council, Inc. Before ERDA in re [NRC] Licensing of ERDA's Projects (July 28, 1975).

34. Letter of November 3, 1975, from R. Tenney Johnson, ERDA General Counsel, to Messrs. R. Cotton and T. Lash, NRDC.

35. *Ibid.*

36. Complaint for Declaratory Judgment and Injunctive Relief, Civil Action File No. 16-1691, U.S. District Court for the District of Columbia, *NRDC et al. v. Seamans et al.,* Sept. 9, 1976.

37. e.g., letter of September 14, 1976, from Peter Strauss, NRC General Counsel to The Honorable Abraham Ribicoff, Chairman, Committee on Government Operations, U.S. Senate.

38. *Ibid.*

39. *Ibid.*

40. *Ibid.*

41. *Ibid.*

42. Letter of November 12, 1975, from Abraham Ribicoff and Henry M. Jackson to Dr. Robert Seamans, Administrator, ERDA.

43. See, e.g., Brief for NRDC (July 28, 1975) (See note 33).

44. See pp. 45–46.

45. President Ford, *Statement on Nuclear Policy,* the White House, October 28, 1976.

46. 42 U.S.C. § 2021 (1970).

47. 42 U.S.C. § 2021(j) (1970).

48. March hearings, 206–207.

49. 42 U.S.C. § 5842(4) (Supp. IV, 1974).

50. From discussions with NRC personnel.

51. Interview with Richard Cunningham, NRC, in Bethesda, Md., June 1976.

52. Appendix F of 10 C.F.R. 50, which defines HL waste, was adopted in 1971; the Energy Reorganization Act was passed in 1974.

53. See p. 52.

54. 18 U.S.C. §§ 831–835 (1970), 49 U.S.C. 1655(c) (4) (Supp. V, 1975); 46 U.S.C. § 170 (1970), 49 U.S.C. 1655(b) (Supp. V, 1975); 49 U.S.C. §§ 1421–1430 and 1472(h) (1970), 49 U.S.C. 1655(c) and (d) (Supp. V, 1975); 49 U.S.C. 1801–1812 (Supp. V, 1975).

55. Memorandum of Understanding between DOT and AEC, 38 *Fed. Reg.* 10437 (1973).

56. *Ibid.*

57. 10 C.F.R. § 71.

58. 10 C.F.R. § 71.12 (1975).

59. See copy in text corresponding with footnote 19.

60. From discussions with NRC personnel.

61. Address by D. A. Nussbaumer, NRC, Denver Symposium, July 14, 1976 (see note 20).

62. 42 U.S.C. §§ 2201(b), (c), (o), 2271–2282 (1970).

63. See, e.g., 10 C.F.R. §§ 30–52, 40.62, 70.55; 10 C.F.R. §§ 19.30, 20.601, 30.61–30.63, 40.71, 40.81, 55.50.

64. ERDA Procurement Instructions Subpart 9-7.50, 9-7.5004-12.

65. See footnote 54 particularly 18 U.S.C. §§ 835 (1970) and 49 U.S.C. 1808–1810 (Supp. V, 1975).

66. 46 U.S.C. § 170 (1970).

67. 46 C.F.R. § 146.02–8(b) (1973).

68. *California Assembly Bill No. 2822,* June 3, 1976.

69. Letter from Governor William G. Milliken to Robert C. Seamans, Administrator, ERDA, July 8, 1976.

70. *Conn. Public Act No. 76-321* (2 *Nuc. Reg. Rep.* Paragraph 20,031 (June 18, 1976)); see also *Miss. Laws of 1976* Chapter 469 (2 *Nuc. Reg. Rep.* Report Letter No. 52 (July 23, 1976)).

71. U.S. Const. art. I, § 8, cl. 3.

72. U.S. Const. art. VI, cl. 2.

73. See 42 U.S.C. § 2021(k) (1970); see generally *Hearings on Federal–State Relationships in the Atomic Energy Field Before the Joint Committee on Atomic Energy,* 86th Cong., 1st Sess. (1959).

74. 447 F.2d 1143 (8th Cir. 1971), *aff'd without opinion,* 405 U.S. 1035 (1972).

75. See *N. Calif. Ass'n to Preserve Bodega Head and Harbor, Inc. v. Pub. Util. Comm'n,* 61 Cal 2d 126, 390 P.2d 200, 37 Cal. Rptr. 432 (1964).

76. See footnote 21 and corresponding copy in text.

77. This project involves the development of an extremely low frequency (ELF) communications system designed to provide military communications for U.S. strategic forces, particularly submarines.

78. See T. Lash, J. Bryson, and R. Cotton, *Citizens' Guide: The National Debate on the Handling of Radioactive Wastes from Nuclear Power Plants,* 34–41, (2d printing 1975).

79. Letter of October 7, 1976, from State Legislature Energy Committee of New Mexico to ERDA.

80. 33 U.S.C. §§ 1401–1444 (Supp. V, 1975).

81. 33 U.S.C. § 1412.

82. *Ibid.*

83. November hearings, 49.

84. May hearings, Vol. 3, 74.

85. 33 U.S.C. §§ 1251–1376 (Supp. V, 1975).

86. 33 U.S.C. § 1311(f).

87. *Train v. Colorado Public Interest Group,* 44 U.S.L.W. 4717 (U.S. June 1, 1976).

88. Convention on the High Seas, April 19, 1958 (1962), 2 U.S.T. 2312, T.I.A.S. No. 5200, 450 U.N.T.S. 82.

89. Convention on the High Seas, art. II.

90. Convention on the High Seas, art. XXIV.

91. Convention on the High Seas, art. XXV, para. 1.

92. Convention on the High Seas, para. 2.

93. Convention on the Prevention of Marine Pollution by Dumping of Wastes and Other Matter, December 29, 1972 (1975), T.I.A.S. No. 8165; for text see 11 I.L.M. 1291 (1972).

94. Convention on the Prevention of Marine Pollution, art. IV, para 1.

95. Convention on the Prevention of Marine Pollution, art. VI.

96. Convention on the Prevention of Marine Pollution, art. VII.

97. Antarctic Treaty, December 1, 1959 (1961), 1 U.S.T. 794, T.I.A.S. No. 4780, 402 U.N.T.S. 71.

98. Antarctic Treaty, art. V.

99. Antarctic Treaty, art. VII.

100. Antarctic Treaty, art. X.

101. Antarctic Treaty, art. XII.

102. See generally D. Deese, "Law of the Sea and High Level Radioactive Waste Disposal: A Potential Geologic Option Under the Deep Seabed?" March 1976 (unpublished thesis for Fletcher School of Law and Diplomacy, Tufts University).

103. President Ford, *Statement on Nuclear Policy,* the White House, October 28, 1976.

104. 42 U.S.C. §§ 5814(c), 5813(1) (Supp. IV, 1974); see also November hearings 27, 31.

105. See *S. Rep. No. 1252,* 93d Cong., 2d Sess. 34–35 (1974).

106. Address by Dr. Carl W. Kuhlmann, Assistant Director for Waste Management, Division of Nuclear Fuel Cycle and Production, ERDA, Denver Symposium, July 13, 1976 (see note 20).

107. *Ibid.*

108. ERDA Program Implementation Document, *ERDA's Program for Management of Radioactive Waste from Commercial Nuclear Power Reactors*, summary at 16 (1976).

109. 42 U.S.C. § 2014(x) (1970).

110. Federal Energy Resources Council, *Status Report on the Management of Commercial Radioactive Nuclear Wastes* (1976).

111. National Environmental Policy Act of 1969, 42 U.S.C. §§ 4321–4347 (1970).

112. NEPA, § 4332(c).

113. *Ibid.*

114. *Ibid.*

115. 42 U.S.C. § 1857h-7 (1970); 42 U.S.C. § 4344(3) (1970).

116. *Natural Resources Defense Council, Inv. v. NRC,* Nos. 75-4276 and 75-4278 (2d Cir., May 26, 1976).

117. *Natural Resources Defense Council v. NRC,* Nos. 74-1385 and 74-1586 (D.C. Cir., July 21, 1976).

118. See *Federal Register,* August 16, 1976, 34710.

119. *Ibid.* at 34707.

120. See *Federal Register,* October 18, 1976, 45849.

121. *Ibid.*

122. Petition of the Natural Resources Defense Council, Inc. Before NRC in re Safety Determination Regarding Disposal of High-Level Radioactive Waste (November 8, 1976).

123. NRDC complaint, September 9, 1976 (see footnote 36), 31, para. 93.

124. Letter of October 27, 1976, from Frank Baranowski, Division of Nuclear Fuel Cycle and Production, ERDA, to Mason Willrich.

125. See *Memorandum of Points and Authorities in Support of NRDC's Petition for Rulemaking and Request for a Programmatic Environmental Impact Statement Regarding NRC Licensing of Disposal of Low-Level Radioactive Wastes* (August 10, 1976).

126. May hearings, Vol. 2, 3–4.

127. President Carter, *Statement on Nuclear Power Policy,* the White House, April 7, 1977.

Chapter 4

Institutional Alternatives

THERE ALREADY EXISTS IN THE UNITED STATES a major radioactive waste problem, which the nation is irreversibly committed to dealing with in some fashion, now and in the future. The technical parameters and size of the problem will change, and it will expand more or less rapidly with continued use of nuclear fuel for electric power production. Radioactive waste has not yet caused substantial and irreparable injury to man or to the environment, although costly mistakes have occurred. We are still free to manage the problem in a variety of ways. Our management options are far broader than the usual shopping list of "technical alternatives." No irreversible institutional commitments have been made. And yet, as Chapter 2 and 3 show, institutions which will be adequate for safe management of radioactive waste in the future are not now in place.

This chapter, therefore, discusses alternatives available for modifying old and developing new institutional arrangements that will be needed to provide reasonable assurance of the safe management of post-fission radioactive waste in the United States in the decades ahead. It is important to emphasize that institutions capable of managing changing problems in a dynamic world must evolve. No organizational arrangement will be appropriate in every situation and for all time. In exploring institutional alternatives for radioactive waste management and regulation, therefore, we are seeking practical arrangements for the next few decades, not a utopian model for the lifetime of the potential radiological hazard.

Policy Goals

Stated objectives of public policy are often vague or general. High public officials may declare that the overarching goal of U.S. radioactive waste policy is ''safe management.'' Scientists may state that our objective must be ''isolation'' of waste from the biosphere for as long as necessary to prevent ''harmful effects'' to man and the environment. However, ''safety'' is not easily defined. ''Isolation'' is a shorthand way of expressing a complicated and relative concept. And what is deemed to constitute ''harm'' to man or environment is a judgment based on subjective values as well as objective facts.

Despite their vagueness, policy goals may be important guides to action. Lack of content does not necessarily indicate the absence of candor. Ultimately, goals are defined through a process of political consensus building. The conditions and content of this consensus change over time. Our goals thus embody shared perceptions of what is an appropriate balance of benefits, costs, and risks.

Discussion of future goals of radioactive waste policy is most usefully organized into two types of concerns: technological and institutional.

Technological Concerns

The technological goals of radioactive waste management will be relatively straightforward, though not easily attained. The basic goal is familiar: sufficient isolation of radioactive waste from the biosphere for long enough to guarantee that the risk of harm to man and to the environment is acceptably low. This is existing policy for permanent disposition of waste. But the specific meaning of terms such as ''sufficient,'' ''enough,'' and ''acceptably low'' are issues which have yet to be adequately addressed in the decision-making process. The goal stated is a sliding one which each of us may judge differently. Therefore, the goal could be met by a spectrum of actions ranging from general, delayed release of the waste material into the environment to complete containment of every nuclide virtually forever. The goal, as stated, is meaningless operationally. The definition of limits for such judgemental terms is a basic step in the decision process by which we will select radioactive waste technologies and the institutions necessary to deploy them.

In the future, it is important to apply comparable technological goals to short-term and long-term waste management. Thus, the likelihood of accidental leaks during waste handling operations should be reduced to very low levels, and there should be appropriate safeguards against malevolent acts at all waste management facilities. Similarly, the desired isolation for the long term may be achieved by (1) assured initial emplacement, (2) reliable containment, and (3) selection of the best practicable geologic formations.

Uncertainties must be identified and effort channeled into areas where uncertainties can be reduced. For instance, the prediction of geologic upheavals may rest on some largely irreducible sources of uncertainty. But significant improvements in the ability to estimate the *consequences* of such an event may be possible by near-term scientific and technological investigation, and such improvements may be very helpful in determining the efficacy of a particular geologic isolation system.

Institutional Concerns

Technological goals are not self-implementing. Institutional mechanisms must be developed to achieve them. The evolving nature of the radioactive waste problem demands institutions which are dynamic. Yet the long-term character of the potential hazard requires unprecedented endurance.

Then upon what or whom should we rely to guarantee that radioactive waste will be isolated from the biosphere for the duration of its potential hazard? It may appear that we confront a clear-cut choice of relying either on nature (geologic disposal) or man (permanent storage). Of course, the real choices are among various combinations of man-made and natural mechanisms, but the emphasis in the combination we choose may turn out to be very important in the long term.

Current policy favors permanent geologic disposal with some reasonable possibility for short-term reversibility. An alternative policy would be reliance on technological systems requiring virtually perpetual surveillance and some degree of active management.

Those favoring primary reliance upon nature may argue that, if the past is any indicator of the future, we can expect political instability and violent social change. Men are fallible and apt to act malevolently. Drastic events resulting in the destruction of society as we know it are possibilities which cannot be ruled out. The future course of human history is unpredictable and reliance upon future generations to deal with the waste problem would be irresponsible. In such circumstances, we owe future generations a guarantee of safety from the deadly waste generated by our nuclear endeavors. Therefore, this generation has the responsibility to commit permanently our radioactive waste to a natural place for safekeeping. If this permanent disposition sooner or later will become irreversible—and with our present technological ability this seems unavoidable—then so be it. Such reasoning has been influential in developing current U.S. policy.

Although geological isolation has obvious attractions, there may be advantages in depending on man more than nature for the time being. The geologic argument rests on our present ability to accurately predict natural history. If we fear human fallibility in the future, how can we be so sure of the correctness of our present judgments about geological isolation? Man has the capacity to learn and thereby achieve a higher level of technological ability. Man can adapt to changing circumstances. Furthermore, if we rely on man-made systems for waste storage, future generations may be in a position to choose to commit the waste problem to nature if it then appears the wisest course to follow. On the other hand, some future civilization may well resent our nuclear age if it has no escape from perpetual care for our waste. And a post-fission civilization may have grave difficulty in continuing to care for fission wastes centuries after all benefits from waste-generating activities have ceased.

Permanent geologic disposal involves present commitments based on present knowledge and predictions of the geologic history of particular formations thousands of years into the future. Once made, geologic dispositions may be extremely difficult, if not impossible, for future generations to reverse if we guess wrong. Permanent geologic disposal thus relinquishes to nature the future management of radioactive waste.

Perpetual human care involves commitments that future generations can modify, including a basic shift from continued surveillance to permanent geologic disposal. As long as technological containment systems function safely, surveillance alone is required. But the safety of reliance on human care rests on a prediction that the stock of knowledge concerning nuclear matters will continue to develop and accumulate and never be substantially lost throughout the future turbulent course of history.

Furthermore, there is no technological artifact in existence today, or that can be reasonably expected to appear in the future, that is capable of performing its originally intended function for many thousands of years. Deterioration in performance is unavoidable. Therefore, perpetual human care entails periodic renewal and replacement of waste containment systems in addition to the surveillance that will always be necessary. Perpetual care retains for man the waste management burden and imposes that burden on successive generations.

Technological systems, which maintain retrievability, and geologic disposal methods, which do not, may both be vulnerable to natural catastrophes. However, technological systems may be more vulnerable to man-made calamities than geologic disposal.

The motives which underlie a policy which favors permanent disposal of radioactive wastes are ambiguous indeed—confidence in our present mastery of nature, fear of future social collapse, or an expedient desire to elicit public acceptance of nuclear power development. Similarly, the motives underlying a policy favoring surveillance are difficult to discern—faith in the capacity of man to muddle through, fear of future natural disasters, or an expedient desire to do the minimum necessary to accelerate nuclear development.

Turning from this rather apocalyptic view of the future, let us now consider an immediate and mundane question. How are radioactive waste policy goals to be achieved? This basic organizational question will be explored by discussion of the three functional areas into which we have previously divided the radioactive waste problem, namely, waste management, safety regulation, and technology development.

Waste Management

Chapter 2 showed, from a technological viewpoint, the desirability of an integrated radioactive waste management framework, and Chapter 3 described the present fragmented management structure.

Now let us reexamine two questions from an institutional viewpoint. First, is there a need for a centralized and integrated management framework for HL waste? And second, does a comparable need exist regarding TRU waste? Thereafter, we will consider a third question: if an integrated management scheme is necessary, what alternative forms are available?

HL Waste

HL waste remains radioactively dangerous for thousands, perhaps hundreds of thousands of years. Since human and environmental protection from the radiological hazard of this waste is our primary concern, the organization of any workable HL waste management scheme must be dictated largely by the long-term nature of the management problem. Although commercial and military waste management problems differ in the short term, the fundamental problem of permanent disposition will be basically the same for HL waste, regardless of its origin. The present structure of long-term HL waste management reflects this analysis. Both commercial and military HL waste will ultimately wind up in a set of repositories, under the management of one federal agency, ERDA. Relatively few permanent repositories will be required to contain both commercial and miltary waste streams to be generated for the foreseeable future. Site selection is arduous and time consuming. Safety considerations dictate storage of the waste in few rather than many repositories if possible. Good long-term management for commercial waste ensures good management for military waste if management is consolidated. This reasoning indicates the desirability of maintaining the unified aspect of the current structure rather than fragmenting it.

The next issue is whether or not short-term HL waste management operations (temporary storage, treatment, packaging, transportation) should be included in an integrated management framework, along with long-term waste repositories. There are a number of advantages to be gained by including short-term commercial and military HL waste operations in a single management authority with long-term repositories. Waste management procedures as a whole are characterized by a high degree of interdependence. Permanent disposition is deeply affected by treatment, which in turn is affected by waste composition. This type of technological inderdependence is an important rationale for integration in any industry (e.g., the petroleum industry). To the extent that separate managers are responsible for the solution to a technologically integrated problem, separate constraints and motivations will be present. This will increase the likelihood of inconsistent strategies for the solution to elements of the problem. We argue that such inconsistencies heighten the risk of an "incorrect" solution to the problem as a whole.

A division of management responsibility would also introduce undesirable transaction costs and risks. Such a separation would provide an incentive for the short-term manager to pass on the maximum amount of cost and risk to the long-term manager, and vice versa, since the operational decisions would be dictated by separate sets of economic and technical considerations. It may be possible to achieve effective results while maintaining separation of long-term and short-term management. But such effectiveness would depend on the development of a high degree of cooperation via regulatory standards and criteria (e.g., acceptance criteria for HL waste forms). The likelihood of conflict in criteria development and implementation would be increased. Moreover, it would be especially difficult to make trade-offs between short- and long-term risks within a managerial framework in which management responsibility is divided.

On the other hand, an integrated management approach would eliminate the short-term/long-term conflicts which a bifurcated system would encourage. An integrated organization would create incentives for waste management efficiency within established safety criteria; the manager would seek to minimize overall cost rather than to pass on costs and risks. Unified long-term/short-term HL waste management would minimize the danger of self-defeating mistakes, since all decisions would be made through a single authority.

Integration of waste management functions is not, however, a virtue in and of itself. It is an attractive organizational solution which grows out of a situation characterized by technological interdependence, a particular economic incentive structure, and the overriding importance of long-term concerns.

It must be remembered that the advantages of any managerial system are necessarily somewhat short-lived. Streamlined management in 1977 does not assure the safe disposition of radioactive waste one thousand years from now. But the theory behind this discussion suggests a positive direction in which to move from the present fragmented management structure.

The possible disadvantages of a management scheme which would pull short-term commercial and military HL waste management into the same structure that handles permanent disposition must be examined carefully. As for commercial HL waste, the only apparent drawback to short-term/long-term integrated management is that it would disrupt the status quo. But the cost of disruption now would actually be minimal, since the structure that currently exists is only a shell: most commercial HL waste is now contained in spent fuel assemblies in storage basins adjacent to reactors.

There are more substantial disadvantages to be considered in the case of military HL waste. First, it might be objected that a new management authority should not be saddled with the existing inventory of mismanaged military waste. In rebuttal, it may be argued that existing military HL waste is the largest and most difficult problem, and the task of dealing with it cannot be shirked. Second, short-term treatment for military HL waste is quite different from that for commercial HL waste. Nevertheless, the long-term solutions and risks are similar. Third, if short-term military and commercial HL waste management operations are merged into the same authority, the public might blend short-term commercial and military practices in its perception of the overall radioactive waste problem. Although past AEC procedures for the temporary storage of military HL waste have been inadequate, there is no reason to suppose that commercial as well as ERDA waste operations cannot achieve a substantially higher level of performance in the future.

A fourth argument against including short-term military waste operations within an integrated management structure along with commercial HL waste is that to do so would jeopardize national security. Disclosure of detailed information regarding military waste composition, it is said, would reveal too much about our nuclear weapons program. But the evidence we have about military HL waste composition indicates that it is a heterogeneous and undecipherable conglomeration. (It may be that certain TRU waste, rather than the HL waste stream, could be revealing about weapons design and production.) Finally, even if there may be a valid security rationale for keeping waste composition classified, there is no reason why the waste management authority could not receive appropriate security clearance. The manufacture of nuclear warhead components and submarine reactor fuel is, in fact, contracted out to private industries.

A final possible drawback to including military HL waste operations in an integrated framework with commercial waste concerns international solutions to permanent disposition. Of course, no international authority would voluntarily assume responsibility for the U.S. military HL waste problem. However, a management structure within the United States that embraces commercial and military waste would not preclude the U.S. waste management authority from participating in multinational joint ventures concerning commercial waste.

TRU Waste

If unified management of HL waste is desirable, should TRU waste also be included within the same management structure? With regard to permanent disposition, the same sort of reasoning applies to TRU waste as to HL waste. In the very long term, the risks incurred in the management of HL waste and TRU waste will tend to equalize, since the quantities of long-lived nuclides (especially plutonium) in these two streams are of the same order of magnitude.

This indicates that a single management framework should be desirable. Somewhat different final disposition technologies may be used for TRU and HL waste, primarily because of their differing thermal properties. But, in the future, permanent disposition of TRU waste is likely to be quite similar to that of HL waste.

Although parallel economic and technological reasoning tends to show that short-term TRU commercial waste management could be handled more safely and efficiently in an integrated management system, the issue is somewhat problematical. Certain aspects of short-term TRU waste management do differ substantially from HL waste operations. TRU waste requires different treatment (e.g., compaction or incineration). Also, the risks involved in the treatment, packaging, transportation, and temporary storage of TRU waste are somewhat less than for HL waste. This may imply the application of different management concepts.

Furthermore, the inclusion of some military TRU waste poses a security problem. Unauthorized personnel must not discover the isotopic compositions of nuclear materials and shapes of components used in the weapons program. This may be a sufficient reason for excluding certain military TRU wastes from the system in the short term. However, since inclusion of military TRU waste in the management structure promises to yield safety and efficiency benefits, the burden ought to be on those concerned about national security to show specifically why particular wastes should be excluded.

Waste Management and Reprocessing

Having considered the advantages and disadvantages of an integrated management structure for commercial and military waste, it is desirable to carry our discussion a step further back into the nuclear fuel cycle. Let us briefly consider implications of the preceding analysis for the spent fuel reprocessing operation from which HL and some TRU waste emerges.

As previously discussed in Chapter 2, reprocessing recovers plutonium and uranium for purposes of recycling it into light water reactors, into breeder reactors, or (in the military program) for use in nuclear weapons. Reprocessing also generates HL and some TRU waste. However, reprocessing and recycling give rise to certain radiological risks themselves. Although ERDA reprocesses spent fuel for military requirements in the United States, there are no operable plants for reprocessing commercial power reactor fuel.

In these circumstances, the question arises: should ERDA reprocessing for military purposes and/or commercial reprocessing for the nuclear power industry become part of an integrated waste management authority? Some factors indicate that inclusion would be desirable.

First, reprocessing and fuel fabrication, and the temporary storage, treatment, packaging, transportation, and final disposition of wastes from these operations are all interdependent steps. Fuel reprocessing and final waste disposition are arguably the most important operations, all others being merely intermediate. The efficiencies of industrial integration, as argued above, might well justify the inclusion of reprocessing in a management structure along with waste operations.

Second, controlling the chemical composition of the waste streams from the reprocessing plant is important as a method of facilitating the subsequent waste solidification process. This would also suggest the inclusion of commercial reprocessing in a unified management structure.

The third factor transcends the issue of management efficiency. Radiological safety might indicate the recovery via reprocessing of as much plutonium and uranium as possible in order to reduce the long-term hazard. (This point, including the plutonium "growth" effects caused by the presence of transplutonium isotopes in the waste, is discussed at greater length in Chapter 2.) However, pushing the technology to the limit for safety reasons could make reprocessing unprofitable as a private commercial enterprise. If reducing potential long-term radiological hazards is the paramount concern, either the government may absorb reprocessing operations into its functions, or it may subsidize private reprocessors to recover more of the long-lived isotopes from spent fuel than is justified from an economic viewpoint.

Nevertheless, the incorporation of ERDA and commercial reprocessing into an integrated waste management structure might not be politically feasible. The ERDA reprocessing operations upstream from HL waste generation relate intimately and sensitively to nuclear weapon requirements. Furthermore, if reprocessing is to be undertaken by the private sector, the recovery and recycling of plutonium and uranium is an economic activity which would result in net benefits to society as a whole. Assuming there are benefits to be realized, the activity may be more efficiently conducted by private enterprise in a market economy.

As a practical matter, however, large government subsidies will be required in order to bring a private reprocessing industry into being in the United States. Moreover, certain key technologies, including plutonium conversion from nitrate to oxide and waste solidification, remain to be demonstrated by the government.

On balance, it seems reasonable to conclude that waste management safety criteria may increase the costs of reprocessing to the point where this step in the commercial fuel cycle becomes unprofitable. If this occurs, government subsidy or take-over would seem preferable to relaxation of safety requirements and resulting increases in long-term radiological risks. On the other hand, the issue of the appropriate future role of the government in commercial reprocessing depends on many factors in addition to waste management. If it is deemed desirable to establish an integrated radioactive waste management system under government auspices, the launching of such an enterprise need not await resolution of that issue.

Once-Through Fuel Cycle

Since reprocessing of commercial power reactor fuel is being deferred in the U.S., it will be necessary to store large numbers of spent fuel assemblies, possibly for long periods of time. If it is subsequently decided to abandon permanently commercial nuclear fuel reprocessing, the issue of final disposition of spent fuel will arise.

In view of the quantities of unreprocessed commercial spent fuel and the probable storage periods involved, substantial storage capacity will be required in addition to that which will exist at power reactor sites. This interim spent fuel storage capacity should be provided at a few central locations. The question thus arises as to whether this capacity would be part of an overall waste management structure, or whether it would be a separate function of the private sector.

Some central storage capacity already exists at private reprocessing facilities that are not now scheduled to become operable, in particular at the General Electric plant at Morris, Illinois, and at the Allied General Nuclear Services plant at Barnwell, South Carolina. Spent fuel is also being stored at the West Valley site of the Nuclear Fuel Services plant. This spent fuel capacity and additional capacity as required might be a private industry function which is separate from any government waste management structure. However, the private sector might not be willing to provide additional capacity, especially since at some unspecified later date there might be either a reversion to a policy of reprocessing or a requirement for rapid final disposition of spent fuel. Both possibilities create uncertainty as to the length of time for which spent fuel storage services will be required—a parameter which will determine the type of storage technology and the capacity necessary.

Alternatively, a unified waste management organization could take over responsibility for spent fuel assemblies following their removal from storage basins at power reactor sites. The organization could then take over existing spent fuel storage capacity from the private sector and provide itself the additional storage capacity that will be required after existing private capacity has been filled. In view of the technical integration of the steps from interim storage through final disposition of spent fuel in the once-through fuel cycle, the long interim storage period that is probable before any final decision is made concerning the disposition of spent fuel, and the possibility of hesitation on the part of private industry to provide storage services (which would only serve to increase the uncertain climate for the nuclear utilities), it appears appropriate for the government to take over responsibility for spent fuel management and that such management should be an integral part of the unified radioactive waste management structure.

Alternative Management Forms

The most appropriate organizational scheme for post-fission radioactive waste management appears to be consolidated management of all waste streams under a single authority. But regardless of whether management is integrated or diversified, what possible forms could the manager or managers assume? Several alternatives should be considered.

Conceivably, the federal government's involvement in waste management might be reduced, rather than increased. This might entail handing over existing waste management responsibility to private industry (subject to regulation), to state governments, to an international authority, or to some combination thereof. If, however, the role of the federal government is to be expanded, three forms are possible: management by agency, by contract, or by corporation.

We can sketch two scenarios in which the federal government's role would shrink and waste management would become a responsibility of private industry, subject of course to government safety regulation. First, waste might be managed by a private entity which would be a monopoly. This could be either an independent corporation or, in the case of commercial waste, a cooperative venture by utilities. Government economic regulation, as well as radiological safety regulation, would be necessary in order to assure that prices charged for waste management services would bear a reasonable relationship to costs and that monopoly profits would not be realized.

As an alternative scenario, waste management might be imagined as a competitive industry. However, there would be formidable obstacles to the growth and operation of competitive enterprise in this field. The initial capital investment for treatment, transportation, and storage facilities would be high and would involve large technological risks. Private industry would be unlikely to proceed in the face of these costs and risks without significant government subsidization. Decentralization of operations in a competitive market would prevent realization of economies of scale that might be possible within a unified management structure. Several firms might not be able to divide the waste disposal market profitably. Furthermore, as experience indicates, more than a handful of acceptable permanent waste repositories will be difficult to locate. Therefore, radioactive waste management could well be considered to be a "natural monopoly."

Competition tends to force private firms to sell their products at a price equal to the marginal cost of production, and over time the marginal cost of the lowest cost producer tends to become the market price. In such circumstances competitive pressures could encourage the trimming of safety related costs. Thus, private waste management firms would tend to generate significant external costs (especially of a long-term nature). Government involvement in the form of taxation or regulation is generally necessary to force industries to internalize such costs. Indeed, radioactive waste is itself an externality of the nuclear power business, and some form of government intervention in waste management has been accepted as necessary to prevent a socially undesirable amount of the potential externalities from being imposed on society as a whole.

Another possible solution would be to turn the radioactive waste problem over to the state governments. A unified management structure would, by definition, preclude this option, since several state governments with diverse interests would be the managers. But if one prefers decentralized management then this is a possibility, at least with regard to the various land-based disposal options. State management would probably be more responsive to the needs of local citizens. But it seems likely that states would rather not be burdened with the expense and responsibility. Many NRC agreement states are now finding the financial strain and political turmoil associated with even low-level waste regulation and surveillance too onerous.

Responsiveness to local needs may imply a flexibility of waste management strategies between states. This contradicts the federal policy of uniformity as a means of assuring safety. State-by-state management would drastically alter the present federally dominated organization. Whereas today federal preemptive power could ultimately compel states to accept national waste repositories, in a state-managed situation one state might well refuse to receive another's nuclear garbage without perhaps exorbitant compensation. On balance, therefore, state management does not seem to be advisable.

To argue that individual states are inappropriate radioactive waste management entities does not imply that the states should have no influence on waste operations within their borders. On the contrary, state participation in congressional and administrative forums for decision making will be a political necessity, especially with regard to such key matters as site selection. ERDA has recognized this requirement in developing its National Waste Terminal Storage Program plan.

A final means of reducing the federal government's role would be to give waste management responsibilities to an international agency. Although this may eventually be desirable, it is quite doubtful that it will occur in the near future. No international agency is likely to take on the responsibility for the U.S. radioactive waste problem, which is one of the most serious worldwide. However, international safety regulation, as distinguished from managment, is feasible and necessary in certain circumstances. This matter is discussed further below.

The foregoing analysis indicates that it would probably be unacceptable to cut back the scope of the federal government's responsibility for waste management by transferring the task to the private sector, to the states, or to an international authority. The other broad option points toward a more unified, integrated management structure under federal government auspices.

The government might unify waste management in three possible ways: by agency, by contract, or by corporation. In the long term, about the same end may be reached using any of these means. However, the management form chosen in the near future will have an important influence on the quality of future decision making and the possibilities for further institutional evolution. Furthermore, the short-term economic and political benefits to be gained from an efficient and effective radioactive waste management organization are important.

Management by executive agency, ERDA, is the predominant form of the current framework. However, numerous activities are conducted by private industry pursuant to government contracts under agency supervision; others are the exclusive management responsibility of private firms, subject to government safety regulation.

In our form of national government, management by executive agency is a common mode. Agencies are entrusted with responsibility for implementing broad congressional policies such as those embodied in the Atomic Energy, Energy Reorganization, and National Environmental Policy Acts. An agency such as ERDA builds up substantial expertise and has significant fiscal resources at its command. Ideally, management by agency is responsive to the president, the Congress, and the public at large.

An objection to government agency operations regarding radioactive waste is their unbusinesslike character. The forces to which an agency such as ERDA is responsive are primarily political, not economic. With regard to the military waste generated by reprocessing operations, ERDA really has no "customer" to whom it is providing waste services. This means it can offer services which are more or less extensive, subject to the constraints of congressional appropriations and public opinion. If commercial reprocessing is authorized, private reprocessors will be ERDA customers for long-term waste management services. But in order to stay in the waste management business, ERDA would not need to charge commercial reprocessors the full cost of waste services provided. The balance might come from public appropriations. If, however, the reprocessor is forced to pay the full cost of waste management, that cost along with others incurred in reprocessing will eventually be reflected in the price for nuclear-electric service (as opposed to higher taxes). Thus with management by government agency, the costs of commercial waste management services may be distributed to society either through the tax structure or the electric rate structure, or some combination of both.

Management by government contract is another possible mode. Here again, this approach is frequently used in the existing arrangements for ERDA operations. Waste management activities may be carried out by a variety of private contractors on a cost-plus basis, under more or less government supervision. Operations may be as tightly controlled as desired, simply by the terms of the contract. This option offers considerable flexibility in the selection of contractors so as to obtain the best available technical expertise in specialized areas. It has the further advantage of being a procedure which is familiar to both government and industry in the nuclear field. Since management by contract is coupled with agency control, it will share the pros and cons of this form outlined above.

The final possibility for an integrated organizational design for the conduct of radioactive waste management would be to establish an independent federal government corporation for this purpose. There are various precedents for this approach, including numerous municipal waste management corporations at the local level such as the New York State Energy Research and Development Authority (NYSERDA), which may take over HL radioactive waste management from Nuclear Fuel Services in West Valley. At the federal level, there are also examples of federally chartered public corporations, such as TVA, although we know of no existing federal corporation with primarily waste management functions. Furthermore, a similar approach has been proposed in the recent report of the Royal Commission on Environmental Pollution, which recommended the establishment of a national nuclear waste disposal corporation for Great Britain.

A federal waste management corporation could be established as an economically viable entity. The corporation could be self-financing, raising capital through the sale of bonds and charging users appropriate prices for its various waste management services in order to recover its costs, including a reasonable return on invested capital. The corporation would be subject to safety regulation by an independent government agency, such as the NRC. However, it could have authority to set prices for its services without prior approval of an economic regulatory commission, as would be required in the case of a private investor-owned utility. Such economic self-regulation is not unusual for public corporations in monopoly positions.

In the establishment of a federal waste management corporation one key factor would be the composition of the board of directors. It would be important to assure that the board was composed so as to assure responsiveness to external criticism and representation of various relevant interests, including certain government agencies, the private nuclear industry, environmental and other citizens' groups, and the academic community. Moreover, the board should not become a political dumping ground.

The executive structure of the corporation should also be designed so as to assure internal incentives for effectiveness and efficiency. Strong incentives for a high quality effort might be provided through clear-cut responsibility for comprehensive problem solving in waste management operations, and by pay scales which would provide compensation for key executives and employees that would be competitive with comparable positions in private industry. To provide flexibility, the corporation might have the option of contracting outside for activities which could not be performed as skillfully or economically "in house."

A federal corporation may be substantially more independent from political pressure than a government agency. Such independence may promote businesslike operations, but a public corporation that is insulated from immediate political pressure from the executive branch and Congress may also be unresponsive to public criticism. Periodic governmental review can help in this respect, and the outcome is also likely to depend on the quality of appointees to the corporate board of directors and key executive positions.

Financing of a radioactive waste management corporation could present cash flow problems initially. For a number of years, the Department of Defense would be the corporation's principal customer. But payments by this customer would necessarily be dependent upon future appropriations by Congress. Financing would have to be assured in the transitional period until the corporation could develop an independent viability.

As regards wastes from the nuclear power industry, it is important to note that ERDA could retain responsibility for research, development and demonstration and a federal corporation might take responsibility for waste management with ERDA-demonstrated and NRC-licensed technology. Thus the establishment of a new federal waste management corporation need not delay, for example, ERDA demonstrations of waste solidification technology and permanent waste repositories.

It is noteworthy that, drawing upon the analysis of our study as originally submitted to ERDA, Senator Charles Mathias of Maryland introduced legislation—the Nuclear Waste Management Act of 1977—which would establish a Nuclear Waste Authority as an independent executive agency. Senate hearings have been held on this legislation, but as of April 1977 no further action has been taken. While we believe a federal corporation is, on balance, preferable to an executive agency, the case is close, and Senator Mathias' proposal merits serious consideration.

Waste Regulation

We have seen in Chapter 4 that the existing regulatory framework for HL and TRU waste is fragmented between NRC and ERDA, with other agencies and certain states playing more or less important subsidiary roles.

Contraction of or withdrawal from safety regulation of radioactive waste operations seems impractical and unwise. Indeed, existing regulation is sparse and in need of future development. This leaves us with three major questions. Is strong federal preemption in the field of radioactive waste regulation desirable? Is a strongly unified regulatory framework more workable than the current one? Should the regulatory agency actively promote or passively review and approve technological options?

Federal Preemption

The federal government now preempts virtually all regulation of HL and TRU waste from a radiological safety standpoint. Thus federal regulation largely displaces state regulation, whether the latter is more or less stringent. Notwithstanding arguments based on states' rights and sensitivity to local opinion, federal preemption will probably continue to be necessary. Uniformity in waste management policy is highly desirable as long as the tendency is to level up, not down toward some least common denominator. Nevertheless, overly stringent state regulation may unduly restrict the nuclear industry or the national security program. Because the adverse consequences of improper waste management might be so widespread, the radioactive waste problem has a national (and international) scope.

Federal preemptive authority will be required in the selection of sites for permanent waste repositories. The best available sites must be chosen and used. But few states are likely to welcome the idea of receiving the nation's radioactive waste within their borders. The only feasible way to eliminate these obstacles in our federal system is to repose paramount regulatory authority in the federal government, subject to advice from any state affected.

It should be emphasized, however, that the notion of federal preemptive power is an ultimate legal standard, not a day-to-day operating procedure. A high degree of state–federal cooperation should prevail.

Integration of Regulatory Framework

Now we must consider whether the currently fragmented federal regulatory framework ought to be consolidated under a single authority. Integrated regulation is feasible even without the kind of unified waste management suggested above.

First, should commercial and military waste be regulated coextensively? Although the actors involved in each case are different, the long-term nature of the problem is the same. (Short-term treatment of military and commercial HL waste could be significantly different, as noted previously.) In view of past experience, regulation of military waste may be necessary to ensure that ERDA proceeds toward a solution to its very difficult and costly military waste problem. Presently, there are no real regulatory incentives for ERDA to do so.

The safety of military waste operations is presently assured by regulatory authority that is divided between short-term ERDA self-regulation and long-term NRC independent regulation. Technically speaking, permanent disposition is deeply affected by earlier reprocessing, treatment, and packaging. It therefore makes little sense to divide regulatory responsibility along an artificial long-term/short-term line. Such a division invites regulatory conflicts and perhaps irrevocable mistakes. Regulation of short-term operations involving military waste may raise problems of dealing with classified information in some areas. But concern for the public health and safety may at least raise a presumption that national security does not justify exemption of a waste management activity from independent safety regulation.

Is coextensive regulation of HL waste and TRU waste appropriate? Albeit less dangerous than HL waste in the short term, TRU waste does require safety regulation. The question is merely who ought to be responsible for its regulation.

TRU waste emerges from more points in the various post-fission nuclear fuel cycle stages than HL waste. TRU waste comes from mixed-oxide fuel fabrication and the head end of the reprocessing plant, as well as the back end. But the multiplicity of sources does not justify fragmented regulation. Moreover, the number of point sources involved in future industry operations should not itself be a major difficulty in regulation. As long as economies of scale in commercial reprocessing and fuel fabrication are realized, the number of facilities involved will remain quite manageable. Finally, unified regulation would eliminate NRC's current inability to exercise *any* authority over military TRU waste regarding either short- or long-term facilities.

Regulatory Role

A final issue is whether or not the regulator, NRC, should actively participate in the selection of technological options to be implemented in waste management activities. Some would contend that a regulator needs to exercise this function in order to promulgate meaningful rules. But a passive role for the regulator is more appropriate for the independent, adjudicatory nature of the licensing function. The regulator's duty is to set levels of performance to be attained in waste management. The manager then has the incentive to devise the most cost-effective means for reaching these criteria. The regulator must then approve or disapprove proposals for operations.

Waste Research and Development

The final radioactive waste function to be considered is research and development. As of now, ERDA carries on most waste management research and development. It operates its research program largely through contracts with laboratories and firms throughout the country. NRC maintains a comparatively modest level of research and development activities, as does the private sector.

Is primary research and development responsibility properly reposed in ERDA or should it be shifted to another entity? A problem with the present system is that management and research and development functions, as well as substantial regulatory authority, are all found within ERDA. Differentiation of these three functions would serve to increase management incentives and, at the same time, help to assure that research and development efforts are objective.

An effective research and development effort is vital to the success of both management and regulation in the long term. In separating management from research and development, there is a possibility that research and development will be pursued increasingly for its own sake, rather than as a process for improvement of operations. Although we conceive of management, regulation, and research and development as three properly separable functions, at least a small amount of research and development will probably be a crucial element of any organization, regardless of its primary mission.

It may be argued that, in a dynamic industrial setting, it is a mistake to separate research and development from operations. If so, the current division between ERDA and the private energy industry with regard to the development and commercial use of many advanced energy technologies is a much more fundamental mistake than would be the case with a federal corporation for radioactive waste management. However, there are also industries where research and development are not a major activity. The electric utility industry is an important historical example. The utilities themselves do relatively little of their own research and development, which is performed instead by the electrical equipment suppliers or the government. (The Electric Power Research Institute is now, however, performing some research and development for the industry as a whole.)

On the other hand, keeping management and research and development under the same roof (ERDA) may mean either that aggressive management of a current problem is hampered by the prospect of a better solution always on the research and development horizon, or alternatively, that promising research avenues are foreclosed too early by the overwhelming pressure of demonstrations which can lead to premature applications. Moreover, separating waste management from waste research and development means that ERDA research and development in this field can also be more responsive to the requirements of an independent regulatory agency, NRC.

Assuming that a rather clear separation of management, regulation, and research and development is, on balance, desirable, the concerns expressed above can be dealt with in determining an appropriate institutional relationship between ERDA, NRC, a federal waste management authority, and the private sector of nuclear industry in addressing two issues: the establishment of radioactive waste research and development priorities, and the development of a mechanism to assure rapid and effective application of promising results of ERDA research and development programs in this field. These issues, summed up in the word "commercialization," are complex and pervade ERDA's primary mission as provider of energy technology to the nation as a whole.

Just as the institutions now in place are in urgent need of repair, we must frankly recognize that whatever institutional arrangements are adopted in the future, sooner or later they too will prove to be inadequate to meet the challenge then posed by radioactive waste. Thus institutional effectiveness must remain a continuing concern.

Chapter 5

Conclusions and Recommendations

POST-FISSION RADIOACTIVE WASTE is highly toxic for long periods—at least thousands, perhaps hundreds of thousands of years. Whether optimists or pessimists, as we look to the future we share an expectation that the vital imperatives of military security and energy supply will surely result in the creation of rapidly growing volumes of radioactive waste in the United States. But is it right to continue activities which generate radioactive waste when a safe method for permanent disposition has not been fully demonstrated?

An optimist may deny there is a serious problem because time and money will provide technology for a variety of solutions. A pessimist may deny there is a solution because—sometimes, somewhere—man-made or natural cataclysms will inevitably breach any technological containment, and toxic radioactive waste may then spill or seep into the biosphere.

The risks posed by radioactive waste must be viewed in context and balanced against the benefits to be derived from activities which produce the waste and the consequences if those activities were stopped. Our security as a nation appears to rest in part on our nuclear deterrent, and the well-being of every society depends on adequate energy. The world urgently needs practical alternatives to fossil energy, and nuclear fission has been demonstrated to be a practical way to generate electricity.

Based on the analysis in the preceding chapters, our principal conclusions and recommendations regarding radioactive waste management and regulation are:

Conclusions

1. *The safe management of post-fission radioactive waste is already a present necessity and an irreversible long-term commitment.*

A rapidly growing number of spent fuel assemblies is accumulating in temporary storage basins at commercial nuclear power plants, pending a government decision on whether to authorize fuel reprocessing. If commercial reprocessing is not authorized, the spent fuel must be safely managed indefinitely. A relatively small volume of HL waste is being stored temporarily in tanks adjacent to a privately owned reprocessing plant that is now shut down. Commercial TRU waste has been disposed of previously in relatively small amounts at various shallow land burial and ocean dump sites. Ocean dumping of U.S. TRU waste has not taken place since 1970.

However, existing military waste will constitute a much larger and more difficult management problem in the near future than the wastes being generated by the commercial nuclear power industry. A large and still growing amount of military HL waste is being temporarily stored in tanks. No specific plan or schedule for treatment and permanent disposition of this waste has been made public. Rough estimates of the cost of preparing the existing inventory of military HL waste for safe permanent disposition run as high as $20 billion. Large amounts of military TRU waste have already been disposed of, and generation and management of this waste continue on a significant scale.

2. *The basic goals of U.S. radioactive waste policy are unclear.*

To the extent that they exist, policy goals have been largely technologically oriented. The technological criteria for permanent disposition of commercial HL waste are containment and isolation from the biosphere for as long as necessary to prevent it from being or becoming a harmful source of radiation. The basic technological goal regarding TRU waste is undecided, although if commercial reprocessing is authorized, the quantities of plutonium contained in this category of commercial waste will be comparable to the plutonium quantities in HL waste.

The technological criteria for military HL and TRU waste management are unclear. Some critics fear that surface tank storage of already solidified HL waste will become a permanent disposition of existing volumes.

In general, radioactive waste policy contains no explicit criteria to apply in developing institutional arrangements for waste management and regulation. Yet institutional effectiveness is an essential ingredient of safe management.

3. *The existing organization for radioactive waste management is likely to be unworkable if left unchanged.*

The management of commercial HL waste is presently divided between private industry and the federal government. The private sector is responsible for temporary storage, treatment, packaging, and transportation; ERDA is responsible for permanent disposition. Bifurcated responsibility for the series of waste management operations, which are technically and functionally integrated, creates incentives for each sector to pass through to the other as many as possible of the risks and costs. Moreover, with responsibility divided, underlying trade-offs between short- and long-term considerations are difficult to make. The existing structure thus tends to prevent, rather than to facilitate, the efficient management of commercial HL waste.

All management functions for commercial TRU waste have rested with the private sector until now. Proposed NRC regulations, however, would shift the task of permanent disposition to ERDA. Collection, temporary storage, treatment, packaging, and transportation would continue to be managed by the private sector.

ERDA is responsible for all stages in the management of most military HL and TRU wastes. Most operations are conducted for ERDA by private contractors.

In the cases of both commercial and military HL wastes, permanent disposition is authorized only at a federal repository on federal land. State land use regulation of the location of permanent repositories for HL waste is thus ultimately preempted by federal law. Nevertheless, through political means and legal and procedural delays, a state government may effectively oppose attempts by the federal government to establish a federal repository within its borders without state consent.

4. *The existing framework for radioactive waste regulation is likely to be ineffective if left unchanged.*

NRC has primary, comprehensive authority to license commercial HL and TRU waste operations from temporary storage through permanent disposition. However, NRC has relinquished regulatory authority over TRU waste to certain states. The Department of Transportation has concurrent regulatory authority with NRC over the transport of radioactive waste.

NRC has authority to license the permanent disposition of military HL waste but lacks authority to license the temporary storage and treatment of such waste. NRC has no licensing authority over military TRU waste. Therefore, no independent regulatory agency licenses existing military post-fission radioactive waste operations to assure the public health and safety. Nevertheless, military HL and TRU wastes constitute the vast bulk of those in existence, and incidents have occurred which have raised doubts about the safety of ERDA's military waste management operations.

Ocean disposal of HL waste is prohibited by U.S. legislation and international law, and ice-sheet disposal in the Antarctic is prohibited by the Antarctic Treaty. The EPA has concurrent jurisdiction with NRC over ocean dumping of TRU waste, although all such U.S. activity has been suspended.

The federal scheme of regulation basically preempts state safety regulation. State land use regulation may, however, operate to affect the location of all radioactive waste operations, except permanent disposition of HL waste at a federal repository. The location of such a repository is a federal matter which ultimately preempts state law. However, as noted above, political and procedural means are available for state opposition.

Pervading the entire framework for radioactive waste regulation are two further features: (1) EPA is responsible for developing generally applicable environmental standards for radiation protection; and (2) the NEPA requires a particular procedure, including preparation of an environmental impact statement, for major federal actions regarding radioactive waste management and regulation.

Recommendations

What steps should we take to strengthen the capacity of our governmental institutions to deal effectively with the radioactive waste problem? We recommend consideration of the following institutional reforms in order to deal more effectively with post-fission wastes:

1. *U.S. radioactive waste policy goals should be clarified to include institutional criteria.*

Important criteria would be:

strong built-in management incentives;
clear differentiation of management, regulation, and research and development functions;
ample jurisdiction for each functional component to perform all interdependent operations within an area of responsibility; and
adaptability to a changing social environment.

2. *A national Radioactive Waste Authority should be established as a federally chartered public corporation.*

The authority would manage all HL and TRU wastes under U.S. jurisdiction or control. Most importantly, the Authority would be responsible for the management of all spent power reactor fuel in the U.S., following removal from temporary storage basins at power reactor sites.

The authority would be independent of ERDA. It would be governed by a board of directors composed of members drawn from government, nuclear industry, the academic research community, and the general public. Except for ERDA research, development, and demonstration activities, the authority would own all HL and TRU waste facilities in the United States. This would include facilities for temporary storage, treatment and permanent disposition of waste, and any specially constructed waste transport containers. It would also own all facilities for storage, treatment, transport and permanent disposition of spent fuel, with the exception of spent fuel storage basins at commercial power reactor sites. The Authority would take over existing commercial and military waste facilities.

The authority would be self-financing. It would issue bonds and recover the full costs of providing waste management services from its customers. It would be authorized to conduct waste management operations itself or to contract with private industry for the conduct of such operations.

The Radioactive Waste Authority would thus be intended to provide comprehensive, integrated, efficient management of both commercial and military HL and TRU wastes. It is noteworthy that the Royal Commission on Environmental Pollution has made a comparable recommendation for establishment of a national nuclear waste disposal corporation in Great Britain.

3. *With NRC as the primary agency, a comprehensive regulatory framework should be established to assure the safety of all radioactive waste management operations under U.S. jurisdiction or control.*

All HL and TRU waste operations, whether involving commercial or military waste, would be subject to NRC licensing. Licensing would be required of existing (unlicensed) military waste operations, as well as future commercial and military waste operations. Various categories of waste from diverse sources may be regulated differently in appropriate circumstances, but all regulation would occur within a unified framework, headed by NRC. Other interested federal and state agencies would play advisory roles.

4. *ERDA should continue to have primary government responsibility for research, development, and demonstration of radioactive waste technology.*

That responsibility would, however, be separated from management of industrial-scale operations on the one hand, and regulation on the other. ERDA-supported radioactive waste research, development, and demonstration activities would be coordinated with and responsive to the needs of both the NRC and the proposed national Radioactive Waste Authority. ERDA would also have responsibility for research, development and demonstration of spent fuel management technologies, including long-term retrievable storage and permanent disposition of spent fuel.

5. *The U.S. government should propose that an international Radioactive Waste Commission be established under the IAEA.*

International approval would be required for any disposition of HL or TRU wastes which would result in emplacement beyond the limits of national jurisdiction. The IAEA commission would also review and comment on proposals for permanent geologic disposition of HL or TRU waste within national jurisdiction.

Finally, it is important to consider the problems of transition from the existing situation to that which we have proposed. The recommendations are not especially sensitive to future scientific revelations or technological developments. Although some of them may appear to be quite far-reaching—especially those regarding a new structure for radioactive waste management—it is important to recall that they can be implemented in most instances with little impact upon large vested interests.

Since research, development, and demonstration would remain ERDA's task, the Radioactive Waste Authority could be launched without any delay in ERDA's plans for radioactive waste demonstration projects. For example, the establishment of a national Radioactive Waste Authority would be compatible with, and could indeed provide additional impetus for, ERDA demonstration of long-term HL and TRU waste disposition techniques and treatment methods for military HL waste, and also an early ERDA demonstration of HL waste solidification.

There is today no long-term management of radioactive waste, no comprehensive scheme for regulation of such waste, and no commercial reprocessing industry in the United States. We believe that institutions can be developed which will provide reasonable assurance of safe management of radioactive wastes in the United States and elsewhere in the world. There is an opportunity to do so now, but it may well be the last clear chance.

Glossary

actinide series: The series of elements beginning with actinium, atomic number 89, and continuing through lawrencium, atomic number 103, which together occupy one position in the periodic table. The series includes uranium, atomic number 92, and all the man-made transuranium elements. The group is also referred to as the "actinides."

cladding waste: Fuel rods in most nuclear reactors today are made up of fissionable materials clad in a protective alloy sheathing which is relatively resistant to radiation and the physical and chemical conditions that prevail in a reactor core. The spent fuel rods, after removal from the reactor and storage to permit radioactive decay of the short-lived fission products, are removed and in certain fuel cycle systems, are chopped up, and the residues of the fissionable materials are leached out chemically. The remaining residues, principally the now radioactivated cladding material (zirconium alloys, etc.) and insoluble residues of nuclear fuel, fission products, and transuranium nuclides, are left behind as cladding waste, which is a special category of transuranium radioactive waste.

critical: The condition in which a material is undergoing nuclear fission at a self-sustaining rate: the critical mass of a material is the amount that will self-sustain nuclear fission when placed in an optimum arrangement in its present form; the minimum critical mass is the amount of a fissile isotope that will self-sustain nuclear fission when placed in optimum conditions.

curie(Ci): A unit of radioactivity defined as the amount of a radioactive material that has an activity of 3.7×10^{10} disintegrations per second (d/s); millicurie (mCi) = 10^{-3} curie; microcurie (μCi) = 10^{-6} curie; nanocurie (nCi)= 10^{-9} curie; picocurie (pCi)= 10^{-12} curie; femtocurie (fCi)= 10^{-15} curie.

decommissioning: The process of removing a facility or area from operation and decontaminating and/or disposing of it or placing it in a condition of standby with appropriate controls and safeguards.

disposal: The planned release or placement of waste in a manner that precludes recovery.

engineered storage: The storage of radioactive wastes within suitable, sealed containers, in any of a variety of structures especially designed to protect them from water and weather and to help keep them from leakage to the biosphere by accident or by sabotage. They may also provide for extracting heat of radioactive decay from the waste.

fertile material: A material, not itself fissionable by thermal neutrons, which can be converted into a fissile material by irradiation in a reactor. There are two basic fertile materials, uranium-238 and thorium-232. When these fertile materials capture neutrons, they are partially converted into fissile plutonium-239 and uranium-233, respectively.

fissile material: Although sometimes used as a synonym for fissionable material, this term has also acquired a more restricted meaning, namely, any material fissionable by neutrons of all energies, including (and especially) thermal (slow) neutrons as well as fast neutrons: for example, uranium-235 and plutonium-239.

fission: The splitting of a heavy nucleus into two approximately equal parts (which are nuclei of lighter elements), accompanied by the release of a relatively large amount of energy and generally one or more neutrons. Fission can occur spontaneously, but usually is caused by nuclear absorption of gamma rays, neutrons, or other particles.

fission products: The nuclei (fission fragments) formed by the fission of heavy elements, plus the nuclides formed by the fission fragments' radioactive decay.

fissionable material: Commonly used as a synonym for fissile material. The meaning of this term also has been extended to include material that can be fissioned by fast neutrons only, such as uranium-238. Used in reactor operations to mean fuel.

fuel (nuclear, reactor): Fissionable material used as the source of power when placed in a critical arrangment in a nuclear reactor.

fuel cycle: The complete series of steps involved in supplying fuel for nuclear power reactors. It includes mining, refining, the original fabrication of fuel elements, their use in a reactor, chemical processing to recover the fissionable material remaining in the spent fuel, reenrichment of the fuel material, and refabrication into new fuel elements, transportation of materials between these various stages, and management of radioactive waste.

fuel reprocessing: Processing of irradiated (spent) nuclear reactor fuel to recover useful materials as separate products, usually separated into plutonium, uranium, and fission products.

half-life: The time in which half the atoms of a particular radioactive substance disintegrate to another nuclear form. Measured half-lives vary from millionths of a second to billions of years. After a period of time equal to 10 half-lives, the radioactivity of a radionuclide has decreased to 0.1 percent of its original level.

high-level liquid waste: The aqueous waste resulting from the operation of the first-cycle extraction system, or equivalent concentrated wastes from subsequent extraction cycles, or equivalent wastes from a process not using solvent extraction, in a facility for processing irradiated reactor fuels. This is the legal definition used by ERDA. Another definition used at the ERDA Hanford Reservation for its waste is: fluid materials, disposed of by storage in underground tanks which are contaminated by greater than 100 microcuries/milliliter of mixed fission products or more than 2 microcuries/milliliter of cesium-137, strontium-90, or long-lived alpha emitters.

ionizing radiation: Any radiation displacing electrons from atoms or molecules, thereby producing ions. Examples: alpha, beta, gamma radiation, short-wave ultraviolet light. Ionizing radiation may produce severe skin or tissue damage.

isotope: One of two or more atoms with the same atomic number (the same chemical element) but with different atomic weights. An equivalent statement is that the nuclei of isotopes have the same number of protons but different numbers of neutrons. Isotopes usually have very nearly the same chemical properties, but somewhat different physical properties.

long-lived isotope: A radioactive nuclide which decays at such a slow rate that a quantity of it will exist for an extended period; usually radionuclides whose half-life is greater than 3 years.

nuclide: A species of atom having a specific mass, atomic number, and nuclear energy state. These factors determine the other properties of the element, including its radioactivity.

partitioning: The process of separating liquid waste into two or more fractions. In this report, partitioning is used specifically with reference to the removal of certain radioisotopes from the waste in order to facilitate subsequent waste storage and disposal. "Isotope mining" is used to describe the fractionation of waste when radioisotopes are extracted and used in other applications.

plutonium: A heavy, radioactive, man-made, metallic element with atomic number 94. Its most important isotope is fissionable plutonium-239, produced by neutron irradiation of uranium-238. It is used for reactor fuel and in weapons.

rad (acronym for radiation absorbed dose): The basic unit of absorbed dose of ionizing radiation. A dose of one rad means the absorption of 100 ergs of radiation energy per gram of absorbing material.

radiation: The emission and propagation of energy through matter or space by means of electromagnetic disturbances, which display both wave-like and particle-like behavior; in this context the "particles" are known as photons. Also, the energy so propagated. The term has been extended to include streams of fast-moving particles (alpha and beta particles, free neutrons, cosmic radiation, etc.). Nuclear radiation is that emitted from atomic nuclei in various nuclear reactions, including alpha, beta, and gamma radiation and neutrons.

radioactive contamination: Deposition of radioactive material in any place where it may harm persons, spoil experiments, or make products or equipment unsuitable or unsafe for some specific use. The presence of unwanted radioactive matter. Also radioactive material found on the walls of vessels in reprocessing plants or radioactive material that has leaked into a reactor coolant. Often referred to only as "contamination."

radioactivity: The spontaneous decay or disintegration of an unstable atomic nucleus, usually accompanied by the emission of ionizing radiation. *Activity* is a measure of the rate at which a material is emitting nuclear radiations, and is usually given in terms of the number of nuclear disintegrations occurring in a given quantity of material over a unit of time. The standard unit of activity is the curie (Ci), which is equal to 3.7×10^{10} disintegrations per second. The words "activity" and "radioactivity" are often used interchangeably.

radioisotope: A radioactive isotope. An unstable isotope of an element that decays or disintegrates spontaneously, emitting radiation. More than 1,300 natural and artificial radioisotopes have been identified.

rem: A unit of measure for the dose of ionizing radiation that gives the same biological effect as one roentgen of X rays; one rem equals approximately one rad for X, gamma, or beta radiation.

roentgen (abbreviation r): A unit of exposure to ionizing radiation. It is that amount of gamma or X rays required to produce ions carrying 1 electrostatic unit of electrical charge (either positive or negative) in 1 cubic centimeter of dry air under standard conditions. Named after Wilhelm Roentgen, German scientist who discovered X rays in 1895.

special nuclear material (SNM): Plutonium, uranium-233, uranium-235, or uranium enriched to a higher percentage than normal of the 233 or 235 isotopes.

transuranics: Nuclides having an atomic number greater than that of uranium (i.e., greater than 92). The principal transuranium radionuclides of concern in radioactive waste management are tabulated below with their half-lives:

NUCLIDE		HALF-LIFE (YEARS)	PRINCIPAL DECAY MODES
Neptunium	237	2,140,000	alpha
Plutonium	238	86	alpha, spontaneous fission
	239	24,390	alpha, spontaneous fission
	240	6,580	alpha, spontaneous fission
	242	379,000	alpha
Americium	241	458	alpha
	243	7,950	alpha
Curium	245	9,300	alpha
	246	5,500	alpha, spontaneous fission

The transuranium nuclide produced in largest amounts is Pu-239; Am-241 is also produced in significant amounts.

transuranic (TRU) waste: Any waste material measured or assumed to contain more than a specified concentration of transuranic elements. This is the definition used by ERDA. The specified concentration is currently set at 10 nanocuries of alpha emitters per gram of waste. (The Nuclear Regulatory Commission [NRC] has not yet adopted a regulatory definition of transuranic waste.) The 10 nanocuries/gm standard is under scrutiny and may be revised upward or completely redefined.

uranium: A radioactive element with the atomic number 92 and, as found in natural ores, an average atomic weight of approximately 238. The two principal natural isotopes are uranium-235 (0.7 percent of natural uranium), which is fissionable, and uranium-238 (99.3 percent of natural uranium), which is fertile. Natural uranium also includes a minute amount of uranium-234. Uranium is the basic raw material of nuclear energy.

Selected Bibliography

BLOMEKE, J. O. *et al.*, "Managing Radioactive Wastes," *Physics Today* (August 1973), 36–42.

COHEN, B. L., "Environmental Impacts of High Level Radioactive Waste Disposal," in *IEEE Transactions on Nuclear Science,* Vol. new series 23, No. 1 (February 1976), 56–59.

COWAN, G. A., "A Natural Fission Reactor," *Scientific American,* Vol. 235, No. 1, 36–47.

CRANDALL, J. L., and C. A. PORTER, *Economic Comparisons of Acid and Alkaline Waste Systems at SRP,* DPST-74-95-37, Aiken, S.C.: DuPont de Nemours, 1976.

DANCE, K. D., *High-Level Radioactive Waste Management: Past Experience, Future Risks, and Present Decisions,* Report to the Resources and Environmental Division of the Ford Foundation, April 1, 1975.

DEESE, D. A., "Law of the Sea and High-Level Radioactive Waste Disposal: A Potential Geologic Option under the Deep Seabed," unpublished thesis, Fletcher School of Law and Diplomacy, 1976.

GILLETTE, R., "Radiation Spill at Hanford: The Anatomy of an Accident," *Science,* 181 (August 24, 1973), 728.

KNEESE, A., "The Faustian Bargain," *Resources,* Resources for the Future, Inc., 44 (September 1973), 1-5.

KUBO, A. S., "Technology Assessment of High-Level Nuclear Waste Management," unpublished Ph.D. dissertation, MIT, 1973.

KUBO, A. S. and D. J. ROSE, "Disposal of Nuclear Wastes," *Science,* 182 (December 21, 1973), 1205.

LASH, T. R. *et al., Citizen's Guide: The National Debate on the Handling of Radioactive Wastes from Nuclear Power Plants,* Palo Alto: Natural Resources Defense Council, Inc., 1975.

LEWIS, R. S., "The Radioactive Salt Mine," *Bulletin of the Atomic Scientists* (June 1971), 27–34.

MURPHY, A. W. and D. B. LAPIERRE, *Nuclear Moratorium Legislation in the States and the Supremacy Clause: A Case of Express Preemption,* Report for the Atomic Industrial Forum, Inc., New York: Atomic Industrial Forum, 1975.

National Academy of Sciences—National Research Council, *Interim Storage of Solidified High-Level Radioactive Wastes,* Washington, D.C.: National Academy of Sciences, 1974.

———, *Report to the Division of Reactor Development and Technology, U.S. Atomic Energy Commission,* Washington, D.C., National Academy of Sciences, 1966.

———, *The Shallow Land Burial of Low-Level Radioactively Contaminated Solid Waste,* Washington, D.C., National Academy of Sciences, 1976.

National Council on Radiation Protection and Measurements, *Alpha-Emitting Particles in Lungs,* NCRP Report No. 46, 1975.

Natural Resources Defense Council, Inc., *Memorandum of Points and Authorities in Support of the Natural Resources Defense Council's Petition for Rulemaking and Request for a Programmatic Environmental Impact Statement Regarding NRC Licensing of Disposal of Low-Level Radioactive Wastes,* August 9, 1976.

———, *Memorandum of Points and Authorities in Support of NRC Licensing of the ERDA High-Level Waste Storage Facilities Under the Energy Reorganization Act of 1974,* July 28, 1975.

———, *Petition to the U.S. Nuclear Regulatory Commission for Rulemaking,* August 10, 1976.

Note, "Application of the Preemption Doctrine to State Laws Affecting Nuclear Power Plants," 62 *Va. L. Rev.,* 738, 1976.

Nuclear Energy Agency and Organization for Economic Cooperation and Development, *Radioactive Waste Management Practices in Western Europe,* Paris: OECD Publications, 1971.

Nuclear Technology 24 (December 1974).

ROCHLIN, G. I., "Nuclear Waste Disposal: Two Social Criteria," *Science,* Vol. 195, January 7, 1977, 23.

SPETH, G. J. *et al., The Plutonium Recycle Decision: A Report on the Risks of Plutonium Recycle,* for the Natural Resources Defense Council, Inc., Washington, D.C.: NRDC, 1974.

STARR, C., "Social Benefit vs. Technological Risk—What is Our Society Willing to Pay for Safety?," *Science,* 165 (September 19, 1969), 1232–1238.

Union of Concerned Scientists, *The Nuclear Fuel Cycle: A Survey of the Public Health, Environmental and National Security Effects of Nuclear Power*, Cambridge, Mass., MIT Press, 1975.

U.S. Atomic Energy Commission, *Additional High-Level Waste Facilities, Savannah River Plant, Aiken, South Carolina (Final Environmental Statement)*, WASH-1530, 1974.

———, *Calcined Solids Storage Additions, National Reactor Testing Station, Idaho (Final Environmental Statement)*, WASH-1529, 1973.

———, *Environmental Survey of the Nuclear Fuel Cycle*, 1972.

———, *Generic Environmental Statement on the Use of Mixed Oxide Fuel in LWR's (Draft)*, WASH-1327, 4 Vols., 1974.

———, *Management of Commercial High-Level and Transuranium-Contaminated Radioactive Waste (Draft Environmental Statement)*, WASH-1539, 1974.

———, *Plan for the Management of AEC-Generated Radioactive Waste (Draft Environmental Statement)*, WASH-1202, 1973.

U.S. Congress, Joint Committee on Atomic Energy, *Storage and Disposal of Radioactive Waste, Hearings*, 94th Cong., November 19, 1975.

———, Senate. Committee on Government Operations, *Low-Level Nuclear Waste Disposal*, 26th Report, 94th Cong., 2d Sess., 1976.

———, *Low-Level Radioactive Waste Disposal, Hearings*, 94th Cong., February 23, March 12, and April 6, 1976.

U.S. Energy Research and Development Administration, *Alternatives for Managing Wastes from Reactors and Post-Fission Operations in the LWR Fuel Cycle*, ERDA-76-43, 5 Vols., 1976.

———, *Proceedings of the International Symposium on the Management of Wastes from the LWR Fuel Cycle*, Denver, Colo., July 11–16, 1976.

———, *Waste Management Operations, Hanford Reservation, Richland, Washington (Final Environmental Statement)*, ERDA-1538, 2 Vols., 1975.

U.S. Federal Energy Resources Council, *Management of Commercial Radioactive Nuclear Wastes: A Status Report*, May 10, 1976.

U.S. General Accounting Office, *Improvements Needed in the Land Disposal of Radioactive Wastes—A Problem of Centuries*, Red-76-54, 1976.

———, *Isolating High-Level Radioactive Waste from the Environment: Achievements, Problems, and Uncertainties*, Red-75-309, 1974.

———, *Opportunity for AEC to Improve its Procedures for Making Sure That Containers Used for Transporting Radioactive Materials Are Safe*, B-164105, 1973.

———, *Progress and Problems in Programs for Managing High-Level Radioactive Wastes*, B-164052, 1971.

U.S. Nuclear Regulatory Commission, Advisory Committee on Reactor Safeguards, *Interim Report on the Management of Radioactive Wastes*, April 15, 1976.

———, "Environmental Effects of the Uranium Fuel Cycle—General Statements of Policy," *Federal Register*, 41, No. 159, August 16, 1976, 34707–34709.

———, *Final Generic Environmental Statement on the Use of Recycle Plutonium in Mixed Oxide Fuel in Light Water Cooled Reactors: Health, Safety and Environment*, NUREG-0002, 5 Vols., August 1976.

———, *Task Group Report—Goals and Objectives for Nuclear Waste Management (Draft)*, June 7, 1976.

———, *Workshop on Waste Partitioning as an Alternative in the Management of Radioactive Waste, Seattle, Washington, June 8, 1976*, 1976.

WEINBERG, A., "Social Institutions and Nuclear Energy," *Science*, 177 (July 7, 1972), 27–34.

WILLRICH, M., *Energy and World Politics*, published under the auspices of the American Society of International Law, New York: The Free Press, 1975.

Index

Index

D

E

F

G

S

T